Strategies for Minimizing Nitrosamine Formation During Disinfection

Awwa Research Foundation

About the Awwa Research Foundation

The Awwa Research Foundation (AwwaRF) is a member-supported, international, nonprofit organization that sponsors research to enable water utilities, public health agencies, and other professionals to provide safe and affordable drinking water to consumers.

The Foundation's mission is to advance the science of water to improve the quality of life. To achieve this mission, the Foundation sponsors studies on all aspects of drinking water, including supply and resources, treatment, monitoring and analysis, distribution, management, and health effects. Funding for research is provided primarily by subscription payments from approximately 1,000 utilities, consulting firms, and manufacturers in North America and abroad. Additional funding comes from collaborative partnerships with other national and international organizations, allowing for resources to be leveraged, expertise to be shared, and broad-based knowledge to be developed and disseminated. Government funding serves as a third source of research dollars.

From its headquarters in Denver, Colorado, the Foundation's staff directs and supports the efforts of more than 800 volunteers who serve on the board of trustees and various committees. These volunteers represent many facets of the water industry, and contribute their expertise to select and monitor research studies that benefit the entire drinking water community.

The results of research are disseminated through a number of channels, including reports, the Web site, conferences, and periodicals.

For subscribers, the Foundation serves as a cooperative program in which water suppliers unite to pool their resources. By applying Foundation research findings, these water suppliers can save substantial costs and stay on the leading edge of drinking water science and technology. Since its inception, AwwaRF has supplied the water community with more than $300 million in applied research.

More information about the Foundation and how to become a subscriber is available on the Web at **www.awwarf.org.**

Strategies for Minimizing Nitrosamine Formation During Disinfection

Prepared by:
Frank Sacher and **Carsten K. Schmidt**
DVGW-Technologiezentrum Wasser (TZW)
Karlsruher Strasse 84, 76139 Karlsruhe, Germany
and
Changha Lee and **Urs von Gunten**
Eawag, Department for Water Resources and Drinking Water
P.O. Box 611, 8600 Dübendorf, Switzerland

Jointly sponsored by:
Awwa Research Foundation
6666 West Quincy Avenue, Denver, CO 80235-3098
and
DVGW-Technologiezentrum Wasser (TZW)
Karlsruher Strasse 84, 76139 Karlsruhe, Germany

Published by:
Awwa Research Foundation

Distributed by:
American Water Works Association
IWA Publishing

DISCLAIMER

This study was jointly funded by the Awwa Research Foundation (AwwaRF) and DVGW-Technologiezentrum Wasser (TZW). AwwaRF and TZW assume no responsibility for the content of the research study reported in this publication or for the opinions or statements of fact expressed in the report. The mention of trade names for commercial products does not represent or imply the approval or endorsement of AwwaRF or TZW. This report is presented solely for informational purposes.

CONTENTS

TABLES

FIGURES

FOREWORD

The Awwa Research Foundation is a nonprofit corporation that is dedicated to the implementation of a research effort to help utilities respond to regulatory requirements and traditional high-priority concerns of the industry. The research agenda is developed through a process of consultation with subscribers and drinking water professionals. Under the umbrella of a Strategic Research Plan, the Research Advisory Council prioritizes the suggested projects based upon current and future needs, applicability, and past work; the recommendations are forwarded to the Board of Trustees for final selection. The foundation also sponsors research projects through the unsolicited proposal process; the Collaborative Research, Research Applications, and Tailored Collaboration programs; and various joint research efforts with organizations such as the U.S. Environmental Protection Agency, the U.S. Bureau of Reclamation, and the Association of California Water Agencies.

This publication is a result of one of these sponsored studies, and it is hoped that its findings will be applied in communities throughout the world. The following report serves not only as a means of communicating the results of the water industry's centralized research program but also as a tool to enlist the further support of the nonmember utilities and individuals.

Projects are managed closely from their inception to the final report by the foundation's staff and large cadre of volunteers who willingly contribute their time and expertise. The foundation serves a planning and management function and awards contracts to other institutions such as water utilities, universities, and engineering firms. The funding for this research effort comes primarily from the Subscription Program, through which water utilities subscribe to the research program and make an annual payment proportionate to the volume of water they deliver and consultants and manufacturers subscribe based on their annual billings. The program offers a cost-effective and fair method for funding research in the public interest.

A broad spectrum of water supply issues is addressed by the foundation's research agenda: resources, treatment and operations, distribution and storage, water quality and analysis, toxicology, economics, and management. The ultimate purpose of the coordinated effort is to assist water suppliers to provide the highest possible quality of water economically and reliably. The true benefits are realized when the results are implemented at the utility level. The foundation's trustees are pleased to offer this publication as a contribution toward that end.

David E. Rager
Chair, Board of Trustees
Awwa Research Foundation

Robert C. Renner, P.E.
Executive Director
Awwa Research Foundation

ACKNOWLEDGMENTS

The authors of this report want to thank all colleagues at TZW and EAWAG for their commitment to the project: Christine Baus for her support during the literature search and for her valuable advice during the laboratory-scale experiments; Silke Glökler, Carola Graf, Sonja Kwiedor, and Thomas Lustinetz for their excellent analytical work which was the basis of the whole project; Muriel Sona and Birgit Körner for their skills in performing all types of laboratory-scale experiments.

We also want to thank all water utilities participating in this project by providing samples as well as water quality data and additional information on their supplies: Philadelphia Water Department (PWD), Long Beach Water Department (LBWD), Charleston Water System, Brick Township Municipal Utilities Authority (BrickMUA), Cedar Rapids Water Department, Sydney Water, Stadtwerke Duisburg AG, Stadtwerke Düsseldorf AG, and Westfälische Wasser- und Umweltanalytik GmbH (WWU). We want to thank especially those colleagues at the waterworks who supported the sampling campaigns by their personal dedication to the project and who answered patently our endless questions.

Special thanks go to the member of the Project Advisory Committee (PAC) for their permanent support of the project: Alexa Obolensky from Philadelphia Water Department (PWD); William Mitch from Yale University; and Andrzej Wilczak from San Francisco Public Utilities Commission (SFPUC). Their comments were always highly appreciated and really brought the project forward.

Last but definitely not least we have to thank AwwaRF project manager Alice Fulmer. She was always there when we needed any administrative or scientific support but never tried to impair our liberties in research. It was really a pleasure to work with her.

EXECUTIVE SUMMARY

BACKGROUND

Nitrosamines are among the most toxic and carcinogenic chemical compounds known. For most of them, carcinogenic effects have been observed already at concentration levels in the low ng/L range. By the U.S. Environmental Protection Agency (EPA), NDMA is classified as "probable human carcinogen" and a 10^{-6} cancer risk level in drinking water at 0.7 ng/L has been determined. In addition to NDMA, the EPA IRIS database includes eight other aliphatic nitrosamines with significant health concerns and 10^{-6} cancer risk levels in the low ng/L range.

Recently, N-nitrosodimethylamine (NDMA) has been identified as a possible disinfection by-product and has been detected in water treatment plants in Canada and in the U.S. using chloramine as disinfectant. Concentration levels found in the finished waters range from few ng/L up to more than 100 ng/L. Previous studies provide first indications concerning possible key factors promoting the formation of nitrosamines during drinking water production, i.e. chloramination or chlorination, long contact times for disinfection, presence of cationic polymers, and surface water sources impacted by waste waters.

RESEARCH APPROACH

There is clear evidence that NDMA is formed during chloramination of waters. With regard to the toxicological properties of NDMA there is a strong need for strategies and methods to minimize NDMA levels in drinking waters. Furthermore, it is necessary to extend the research to other nitrosamines, as other nitrogen-containing compounds are present in natural waters that undergo disinfection and might react with monochloramine by forming other nitrosamines than NDMA.

Thus the main objective of this research project was the development of strategies for minimizing nitrosamine levels in drinking waters. The project did not only focus on formation of the most prominent nitrosamine NDMA but also covered other aliphatic nitrosamines with significant health concern as N-nitrosodiethylamine (NDEA), N-nitrosomethylethylamine (NEMA), N-nitrosodipropylamine (NDPA), N-nitrosodibutylamine (NDBA), N-nitrosopiperidine (NPIP), N-nitrosopyrrolidine (NPYR), and N-nitrosomorpholine (NMOR).

For this purpose different approaches were followed. First, information on mechanisms of nitrosamine formation was collected in order to identify options for optimizing operating conditions for disinfection with regard to nitrosamine formation but without compromising disinfection efficiency. Furthermore, significant efforts to identify or at least characterize possible precursors of nitrosamines were undertaken and options for removal of precursor compounds prior to disinfection were studied. Finally, possibilities for nitrosamine removal were evaluated. For all of these approaches (optimization of operating conditions, removal of precursors, removal of nitrosamines), an extensive literature search was performed compiling the present knowledge on formation of nitrosamines during treatment of wastewaters and drinking waters, occurrence of possible precursors in the aquatic environment and removal of nitrosamines during treatment. Additional laboratory-scale experiments were performed in order to fill knowledge gaps identified during the literature search. For this, tests for the determination of the nitrosamine formation potential (NFP) were performed. With these NFP tests, water samples could be characterized with regard to their potential to form nitrosamines. The levels of nitrosamines formed in these tests was used to compare the potential of different waters to form nitrosamines during chloramination as the amount of nitrosamines formed characterizes the quantity and potency of precursors present in the sample under investigation. However, the NFP test was not thought to predict concentration levels of nitrosamines formed during disinfection in waterworks or in distribution systems.

The conclusions from the literature search and the laboratory-scale experiments were then verified by full-scale tests in water utilities in U.S., Australia and Germany. Taking into account the outcome of the literature search as well as the results of the laboratory-based experiments with water samples from water utilities, options and strategies for minimizing nitrosamine formation during disinfection of drinking water were derived.

FORMATION OF NITROSAMINES

The chemistry of nitrosamine formation in general is very complex. Many reactions can occur simultaneously and the products depend on the exact reaction conditions, including concentrations of reactants, catalysts and inhibitors, and the relevance of other reactions which can compete for the nitrosating agents. Furthermore, most of the published work focuses on NDMA formation, while formation of other nitrosamines has attracted only little attention in the past. However, most of the principles directing NDMA formation should, at least on a qualitative basis, also be valid for the formation of other nitrosamines.

According to the literature studied, there are several pathways by which nitrosamines can be formed: (a) Nitrosation of nitrogen-containing compounds by nitrosating agents, (b) Reaction of monochloramine with aliphatic amines (e.g. dimethylamine) to unsymmetrical dimethylhydrazine (UDMH) and subsequent oxidation to nitrosamines, and (c) Reaction of monochloramine with aliphatic amines to chlorinated UDMH and subsequent oxidation to nitrosamines. Possible nitrosating agents for pathway (a) are NO^+ (nitrosonium ion), $H_2NO_2^+$ (nitrous acidium ion), HNO_2 (nitrous acid), N_2O_3 (nitrous anhydride), N_2O_4 (dinitrogen tetroxide), XNO (nitrosyl halide; X = Cl, Br), and NCSNO (nitrosyl thiocyanate). Possible nitrogen-containing precursors for N-nitrosamines are secondary, tertiary, and quaternary aliphatic and aromatic amines, N,N-dialkylhydrazides, N-alkylamides, N-alkylureas, 1,1-dialkylureas, 1,3-dialkylureas, 1,3-dialkylphenylureas, tetraalkylureas, N-alkylcarbamates, and 1,1-dialkylthioureas. Additionally, organic compounds having structural elements of the above mentioned compounds can act as precursors for nitrosamine formation, e.g. pesticides, cosmetics, sunscreen agents, pharmaceuticals, or even natural organic matter (NOM). Nitrosation might be enhanced via catalysis by metal salts (e.g. Cu, Fe, Zn), by nucleophilic anions (e.g. thiocyanate and to a smaller extend halides), formaldehyde and chloral, phenols, and fulvic acid. Reducing agents such as ascorbic acid, on the other hand, can minimize or even prevent nitrosamine formation by reducing nitrite to nitric oxide.

The reaction pathway which seems to be especially relevant for NDMA formation during disinfection of waste and drinking waters with chlorine and chloramine involves a reaction of monochloramine with dimethylamine to either UDMH or chlorinated UDMH and subsequent oxidation to NDMA. The proposed mechanisms account for the fact that NDMA formation has been observed during disinfection with chlorine and chloramines even in the absence of nitrite. If NDMA is formed via these pathways, precursors are restricted to dimethylamine and compounds containing dimethylamine functional groups. After chloramination of wastewater treatment plant effluents, NDMA was found besides other nitrosamines like nitrosodiethylamine, N-nitrosopyrrolidine, and N-nitrosopiperidine.

To complete the information available in literature, laboratory-scale experiments were performed. Using the NFP test, conversion rates of aliphatic amines into the corresponding nitrosamines during chloramination were determined. It could be demonstrated that the percentage of conversion based on the molar concentrations is generally rather low, ranging from 0.02 to 2.7%, depending on the amine tested and its concentration. In addition, catalytic effects were studied in more detail. However, formation of nitrosamines during chloramination was in neither case substantially enhanced by the anions and cations tested. NDMA and NPYR formation rather turned out to be significantly lowered in the presence of thiocyanate and nitrite, implying that both anions might react faster with chloramine than precursor compounds relevant for nitrosamine formation.

Based on these findings, nitrosamine formation by chloramination from secondary amines has to be regarded as rather inefficient and neither matrix nor catalytic effects seem to improve conversion rates substantially.

From a series of chloramination and chlorination experiments it became obvious that chlorination with hypochlorite is much less critical with regard to NDMA formation than chloramination. Thus, the formation of NDMA via the dinitrogen tetroxide pathway during chlorination described in literature seems to be much less significant compared to NDMA formation during chloramination.

Looking at the formation of the different nitrosamines in the NFP tests, it can be concluded that the formation of other nitrosamines than NDMA is almost negligible compared to the formation of NDMA. In all NFP tests, the levels of NDEA, NEMA, NDPA, NDBA, NPIP, NPYR, and NMOR were one or two orders of magnitude lower than the concentrations of NDMA and in most experiments, no other nitrosamine than NDMA could be detected.

OCCURRENCE OF POSSIBLE PRECURSORS

Organic nitrogen-containing compounds that might form nitrosamines during chloramination are numerous. Chloramination of surface waters in laboratory-based experiments under conditions that are comparable to those found in drinking water treatment indicated that the concentration of NDMA produced from naturally occurring precursors usually will be less than 0.1 nM (7.4 ng/L) and thus NOM as major precursor for nitrosamine formation during chloramination can be ruled out.

Polyelectrolytes used in water treatment as primary coagulants or as floc builders, are described in literature as important source of NDMA precursors during drinking water treatment. It was shown in several laboratory-based and field studies that the use of polyDADMAC and epi-DMA in the coagulation/flocculation/sedimentation process in conjunction with chlorination causes significant NDMA formation.

Anion exchange resins which are commonly used in various ion-exchange processes in drinking water treatment consist primarily of a polymer matrix and a quaternary amine function as ion-active group. Several studies have reported on detectable NDMA levels in waters exposed to certain anion exchange resins and it was concluded that most likely the NDMA originated from quaternary amines used to prepare these resins. The origin of the nitrosamines, however, is not clear. There might be either impurities caused by the manufacturing process or nitrosamines are formed during application of the resin to certain types of water, eventually promoted via surface-catalyzed reactions.

However, many nitrogen-containing compounds that might act as precursors for nitrosamine formation during chloramination can be found in the aquatic environment, e.g. in industrial wastewaters, surface waters, and raw waters used for drinking water production. Concentration levels of single compounds, however, are in most cases in the low µg/L level or even below 1 µg/L and, thus, could not completely account for the nitrosamine levels found in laboratory-scale experiments and in finished drinking waters after chloramination taking into consideration available kinetic data on nitrosamine formation.

Thus, additional laboratory-based experiments were performed in order to gain further insight into nature and origin of precursors responsible for NDMA formation during chloramination. The studies should also clarify to what extent other nitrosamines than NDMA are relevant for water suppliers utilizing chloramination. For this, about 100 samples were chloraminated with the NFP test and formed nitrosamines were determined by chemical analysis. Additionally, tested water samples were characterized by a set of physical-chemical parameters. The results from these experiments clearly proved that NDMA was the main nitrosamine formed during chloramination in all waters investigated. NPYR and NDEA were the nitrosamines second and third most often found, however, levels were always at least by an order of magnitude lower than those of NDMA. In surface waters showing an extremely high NDMA formation potential, low levels of NEMA, NDPA and NPIP were also detected occasionally. Typically, these surface waters were also characterized by a high waste water load. On the other hand, surface waters without any waste water portion showed only little potential for NDMA formation during chloramination (maximum NFP of 7.1 ng/L NDMA), confirming again the small importance of natural organic matter for nitrosamine formation during chloramination.

For surface water samples, a good correlation between boron concentration and NDMA formation potential was found, underlining the importance of anthropogenic water constituents for NDMA formation during chloramination. None of the other parameters analyzed in the basic set of physical-chemical parameters showed a comparable good correlation to the NFP.

The groundwater samples under investigation showed rather low NDMA levels in the NFP test when compared to surface waters. Within groundwater samples, the waters being influenced by surface water and those with higher TOC levels seem to be more susceptible to NDMA formation during chloramination than non-influenced waters with lower TOC levels.

Numerous samples were analyzed in parallel for primary and secondary aliphatic amines. It was found that concentrations of amines rarely exceed levels of 200 ng/L per single compound. Based on the conversion rates determined in laboratory-based experiments, concentrations of 200 ng/L amine would be capable of forming about 8.5 ng/L NDMA, 1.4 ng/L NDEA, 5.1 ng/L NPYR, or 0.3 ng/L NMOR. From these calculations it is evident that at least NDMA and NPYR formation during chloramination is mainly based on other precursors than primary or secondary aliphatic amines.

Thus, a set of tertiary amines was selected for chloramination experiments in order to study the influence of various structural elements of tertiary amines located next to the dimethylamine function. The results obtained from these experiments clearly demonstrated that nitrosamine formation is substantially enhanced for some tertiary amine compounds providing certain structural characteristics. Among the substances tested, ranitidine, a pharmaceutical often used against gastritis, showed the highest conversion efficiency of 62.9%. Thus, with the same number of molecules ranitidine forms about 83fold more molecules NDMA than does DMA. Several tertiary amines showed conversion efficiencies comparable to those of dimethylamine and trimethylamine in the range of 0.7 to 3% and some tertiary amines even showed less conversion than dimethylamine. A closer look at the structural characteristics of these compounds revealed two important facts. NDMA formation from a structure with the dimethylamine group next to a carbonyl or a sulfonyl group is typically rather low, while structures with the dimethylamine group being coupled over a methylene group in α-position to an electron-rich, heterocyclic, 5-membered ring system (e.g., furan or thiophene) typically form NDMA with very high conversion rates.

The comprehensive evaluation of the data implies that water utilities using surface water under the influence of wastewater are at a higher risk for nitrosamine formation during chloramination than those making use of non-impacted water sources.

REMOVAL OF POSSIBLE PRECURSORS

Options for removal of possible precursor compounds for nitrosamine formation during chloramination were studied by literature search as well as by laboratory-scale experiments.

According to literature, biodegradation, e.g. during underground passage, is an option for removal of some nitrogen-containing compounds whereas others turned out to be rather persistent. Most of the aliphatic amines seem to be efficiently removed by biodegradation processes under aerobic and anaerobic conditions. In order to investigate the influence of biofiltration processes on the removal of nitrosamine precursor compounds present in wastewater-impacted surface waters, microbial degradation was studied in the laboratory using a biologically active test filter. This simulation method gives basic information about the biodegradability of NDMA precursor compounds excluding sorption effects to sediment or soil. By applying the NFP test to surface waters treated in the test filter, it was found that NDMA precursor compounds tend to be rather persistent, as concentrations decreased only slowly with roughly 70-90% of the initial amount persisting even after one month. Most interestingly, no major differences in degradation rates were noticed between different surface waters, even though NDMA precursor compounds were of different initial concentration and origin.

Flocculation and sedimentation are described in literature as only poorly removing DON as well as single aliphatic amines. Nevertheless, the impact of flocculation on the removal of nitrosamine precursor compounds was investigated in laboratory-scale tests. The results clearly demonstrated flocculation cannot be regarded as efficient treatment option for the removal of precursors for nitrosamine formation during disinfection. Observed reduction rates for formation of NDMA, NDEA, and NPYR after flocculation with either iron or aluminum salts were below 10%.

According to literature, aliphatic amines are not effectively oxidized by ozone but do react rather fast with chlorine. It is further described that pre-chlorination can lower the NDMA formation potential substantially. Consequently, the removal of nitrosamine precursors by oxidation processes was investigated in more detail. Second-order rate constants for reactions of selected NDMA precursors with ozone and chlorine dioxide were determined in order to assess the degradation kinetics during the oxidation processes. It was found that DMA as an important NDMA precursor is possibly produced from the oxidation of tertiary amines with a DMA functional group. The oxidation of NDMA precursors with chlorine dioxide tended to yield a higher concentration of DMA than with ozone. In further experiments, the change in NDMA formation potential after oxidation of possible precursors with ozone and chlorine dioxide was investigated (other nitrosamines have not been investigated in those tests). It was found that the oxidation of NDMA precursors with ozone and chlorine dioxide significantly reduced the NDMA formation potential. To investigate the effect of pre-oxidation on the NDMA formation potential of natural waters, experiments were performed using natural surface water samples. As expected from the results with model NDMA precursors, the pre-oxidation significantly reduced the NDMA formation potential of natural waters. The effectiveness of chlorine dioxide and chlorine for reduction of NDMA formation potential was found to be similar to that of ozone.

Adsorption of possible precursors on activated carbon strongly depends on the chemical structure of the compounds. DON is removed to ca. 50% by activated carbon whereas aliphatic amines are only poorly adsorbed. For studying the removal of precursor compounds for nitrosamine formation by activated carbon adsorption, laboratory-scale batch experiments with powdered activated carbon (PAC) were performed. The results of these experiments clearly indicate that the precursors for nitrosamine formation adsorb quite well onto activated carbon. Already by addition of a PAC dose of 5 mg/L, the NDMA formation potential of the river waters could be decreased by at least a factor of 2 and the same is true for the NMOR and NPYR formation potential. Evaluation of the obtained experimental data according to the equation of Freundlich indicated that NDMA formation in the sample under investigation can be attributed to a single compound or a group of very similar compounds and not to a complex mixture of different compounds.

REMOVAL OF NITROSAMINES

Nitrosamines are rather persistent in the aquatic environment. However, it has been clearly demonstrated in literature that they can be mineralized by microbial communities under defined conditions. This process is relatively slow and levels of nutrients and co-factors seem to be of major importance for successful nitrosamine elimination. Reported half-lives for nitrosamine biodegradation in soil and water are in the range from several days over months to even years. As it was still unclear how nitrosamines like NDMA behave during bank filtration, artificial recharge, or biofiltration units used for drinking water treatment, additional experiments were performed to clarify if such treatment steps might be an option for NDMA removal under field conditions. Using a so-called test filter to simulate biodegradation under aerobic conditions, it was demonstrated that NDMA concentrations decreased quite rapidly, indicating a fast microbial degradation. Compared to NDMA, the concentrations of the other nitrosamines decreased much more slowly. According to the results of these experiments, aerobic biodegradability of the nitrosamines decreases in the order NDMA >> NDBA > NDcHxA > NDEA > NPIP, NDPA, NEMA > NPYR > NMOR.

NDMA and other nitrosamines are not easily removed by conventional treatment technologies. GAC adsorption of nitrosamines is assessed as being rather poor while carbonaceous resins show better adsorption characteristics. Best performance was found for the adsorption of nitrosamines on zeolites. However, the adsorption capacities of those materials for nitrosamines are still low compared to other organic substances. NDMA and other nitrosamines can also not efficiently be removed from water by volatilization. Thus, air stripping is not an option in treating NDMA contaminated water.

Ozonation has no effect on NDMA concentration. The application of Advanced Oxidation Processes is more likely to reduce concentrations of NDMA and other nitrosamines. The determined reaction rate constants, however, are well below those of most other organic substances. Fenton-type reactions, i.e. reactions involving OH radicals formed by an iron(II) catalyzed destruction of hydrogen peroxide, proved to be rather efficient in eliminating NDMA. In additional laboratory-based experiments the rate constants for the reaction of various oxidants with

NDMA were determined. Again, NDMA proved to be strongly resistant to oxidation because of the electron withdrawing nitroso group. The calculated rate constants are very low except for the reaction with $^{\bullet}OH$, which is known as a highly reactive and non-selective oxidant.

Oxidation of NDMA via hydrolysis is strongly dependent on pH. Under acidic conditions (pH < 4.5) an increase in reaction rate is observed though still the absolute value is rather low. Catalytically enhanced hydrolysis, e.g. by addition of zeolite, silica and copper or nickel ions showed a removal efficiency of 20%. Reverse osmosis showed some potential for NDMA removal. Thin film composite membranes yielded a 40% reduction of NDMA concentration.

The most promising technology investigated in literature is the photolysis of NDMA by UV irradiation and some pilot-scale and industrial-scale applications are described in literature. The UV dosage required for NDMA removal of one order of magnitude (approximately 1,000 mJ/cm^2) is ten times higher than that for equivalent virus removal. This means that an effective NDMA removal by UV irradiation leads to a complete disinfection of the treated water.

FULL-SCALE TESTING IN WATER UTILITIES

In order to verify the results from laboratory-scale experiments under real-world conditions, full-scale tests were performed. Samples from five U.S. utilities, one Australian utility, and three German utilities were obtained and analyzed for nitrosamines. Additionally the nitrosamine formation potential was determined for each sample using the NFP test. During these full-scale tests, treatment steps as bankfiltration, artificial groundwater recharge, flocculation, aeration, lime softening, ozonation, chlorination, and activated carbon filtration were evaluated with respect to their efficiency to remove precursors for nitrosamine formation during chloramination.

One observation during the full-scale tests was that the use of groundwater as raw water source seems to be beneficial with regard to nitrosamine formation during chloramination. In most cases, nitrosamine formation in the NFP tests was lower for utilities treating groundwater than for utilities using river water as raw water source.

The results from the full-scale tests clearly confirmed the findings from the laboratory-scale experiments on the aerobic degradation of NDMA. At one utility, NDMA was found to be well eliminated (> 90%) during the aerobic riverbank filtration process. Additionally it was shown that concentration levels of aliphatic amines that might act as nitrosamine precursors are also reduced by the riverbank filtration process, even though a complete removal could not be observed. Average elimination rates were 76% for DMA, 66% for DEA, 52% for pyrrolidine and 80% for morpholine.

Also at another plant, riverbank filtration turned out to be capable of removing precursors of NDMA, NPYR, and NDEA to a large extent, with removal efficiencies for the NDMA formation potential of 93%, for the NPYR FP of > 50%, and for the NDEA FP of > 33%. In this plant, NDMA precursor levels in the raw riverbank filtrate could be further reduced by about 90% via the combination of activated carbon treatment and ozonation. Also the results of these full-scale experiments confirm the results from the laboratory-scale studies where treatment steps based on activated carbon and ozonation were demonstrated to be effective in NDMA precursor removal.

The results from several plants clearly indicated that neither aeration nor lime softening result in a significant removal of precursors for nitrosamine formation during disinfection. Also conventional treatment steps of coagulation, flocculation, sedimentation, and filtration do not significantly reduce the amount of nitrosamine precursors. In contrast, for one utility the use of a coagulant aid was associated with a significant increase in NDMA formation potential.

Further full-scale tests, however, clearly proved that that oxidative treatment steps such as ozonation or chlorination lead to a significant reduction in NFP. In most cases, the removal of nitrosamine precursors turned out to be much more efficient than for natural organic matter in general.

The full-scale tests also confirmed the finding of the laboratory-scale experiments that NDMA is by far the most relevant nitrosamine formed during chloramination. In the original water samples obtained from the water utilities, only NDMA could be detected in the chloraminated water samples. After the NFP test, especially NPYR and in one case NDEA and NEMA were found in addition to NDMA, however, at much lower concentration levels.

STRATEGIES FOR MINIMIZING FORMATION OF NDMA AND OTHER NITROSAMINES DURING DISINFECTION OF DRINKING WATER

From the outcome of the literature search as well as from the results of the laboratory-scale experiments and the full-scale tests at several water utilities, several options for minimizing nitrosamine formation became apparent.

The first option takes into account the raw water source. Most of the precursors for nitrosamine formation during chloramination can be related to anthropogenic sources and thus water utilities using surface waters that are characterized by high wastewater loads are more prone to nitrosamine formation than those using non-impacted water sources for drinking water production. Water utilities that have no alternative to using wastewater-impacted surface water might utilize treatment processes as slow sand filtration, artificial groundwater recharge or riverbank filtration to reduce levels of nitrosamine precursors. Besides the removal of precursors, these treatment steps would also reduce concentration levels of nitrosamines already present in the surface waters.

Optimizing operational parameters of the chloramination process is another possibility for minimizing nitrosamines levels in the disinfected waters. It is obvious that decreasing the disinfectant dose leads to a reduction of nitrosamine levels. Lowering the pH is another option, as nitrosamine formation during chloramination is more efficient at pH values of 8 and above than at a pH of around 7. When changing operational parameters, however, the microbiological safety of the disinfected water as well as possible corrosion problems must always be taken into account. In most cases, the use of chlorine (or hypochlorite) as disinfectant instead of chloramine will also lead to a significantly lower nitrosamine formation. However, it has to be taken into account that chlorine produces other by-products of toxicological relevance and that chlorine has disadvantages with regard to disinfection efficiency and stability in the distribution system. Laboratory-scale experiments indicate that the rate of NDMA formation from the reaction of dichloramine with organic nitrogen-containing precursors is much faster than the analogous reactions involving monochloramine. Thus one strategy to minimize NDMA formation during chloramination might be the production of chloramines in a manner that minimizes the concentration of dichloramine, e.g. by addition of free chlorine before ammonia.

A promising strategy for minimizing nitrosamine formation during chloramination is the removal of precursor compounds prior to disinfection. Microbiological degradation during slow sand filtration, artificial groundwater recharge, or riverbank filtration is an efficient option for removal of nitrosamine precursors whereas treatment steps such as aeration or lime softening have no impact on the nitrosamine formation potential and thus can not be used for minimizing nitrosamine formation during chloramination.

Flocculation and sedimentation also turned out to be inefficient in removal of nitrosamine precursors. In contrast, polyelectrolytes used in water treatment as primary coagulants or as floc builders, are important sources of NDMA precursors. It was shown that especially the use of polyDADMAC and epi-DMA in the coagulation/flocculation/sedimentation process causes significant NDMA formation during chlorination and chloramination and thus the use of these polymer materials should be avoided if the treated water is disinfected with chlorine or chloramine. Furthermore, the utilization of membrane technologies for separation will reduce the need for polymers and thus will eliminate this precursor for nitrosamine formation.

A much better option for minimizing nitrosamine formation is the inclusion of a pre-ozonation or pre-chlorination step in the treatment process. Laboratory-scale experiments as well as the full-scale tests at several water utilities clearly proved that oxidation is an efficient tool for removal of nitrosamine precursors. Observed reduction rates for NDMA formation after pre-ozonation as well as after pre-chlorination amounted to more than 90%. Both treatment steps, however, produce other by-products than nitrosamines and thus always the whole treatment process and all unwanted by-products have to be taken into account. Furthermore, pre-oxidation steps have to be carefully optimized with respect to location, dose of oxidant, contact time, and pH in order to minimize by-product formation and to optimize oxidation efficiency.

When looking at the possible treatment options for removal of nitrosamine precursors, it should always been taken into account that these processes not only efficiently minimize nitrosamine formation during chloramination but remove a lot of other unwanted compounds from the raw water.

Removal of nitrosamines formed during chloramination of drinking waters can not be regarded as primary option. Chloramination in most cases is the final treatment step in a waterworks and thus an additional post-treatment is not feasible. Furthermore, the options for nitrosamine removal are limited as NDMA and other nitrosamines are not easily removed by conventional treatment technologies like flocculation, ozonation or GAC filtration.

CHAPTER 1
INTRODUCTION

BACKGROUND

Nitrosamines, i.e. N-nitroso derivatives of aromatic and aliphatic amines, are among the most toxic and carcinogenic chemical compounds known. For most of them, carcinogenic effects have been observed already at very low concentration levels. Table 1.1 summarizes important nitrosamines. Their major physical-chemical properties are displayed in Table 1.2.

Table 1.1
Important nitrosamines

Nitrosamine	Short form	CAS No.	Molar mass (g/mol)	Chemical structure
N-Nitrosodimethylamine	NDMA	62-75-9	74.08	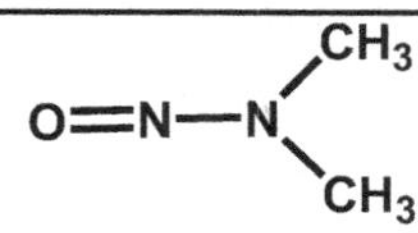
N-Nitrosoethylmethylamine	NEMA	10595-95-6	88.11	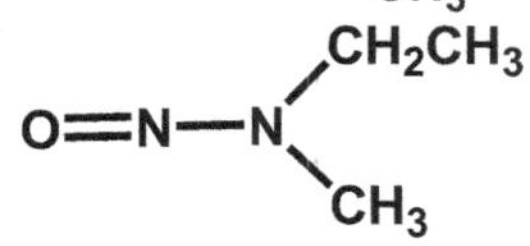
N-Nitrosodiethylamine	NDEA	55-18-5	102.14	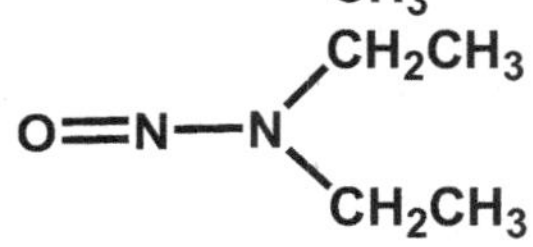
N-Nitrosodi-n-propylamine	NDPA	621-64-7	130.19	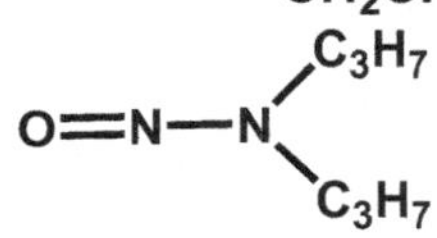
N-Nitrosodi-n-butylamine	NDBA	924-16-3	158.24	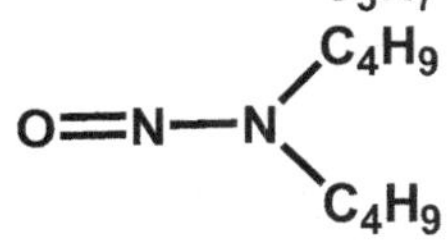
N-Nitrosopiperidine	NPIP	100-75-4	114.15	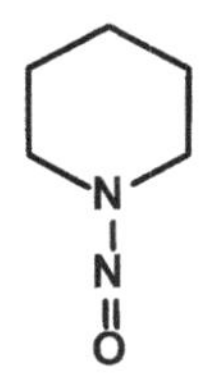
N-Nitrosopyrrolidine	NPYR	930-55-2	100.12	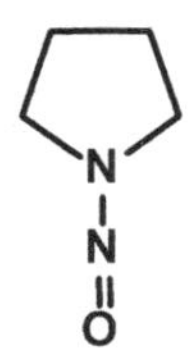
N-Nitrosomorpholine	NMOR	59-89-2	116.12	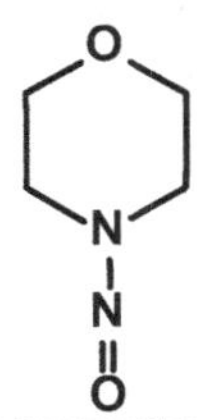

Table 1.2
Physical-chemical properties of selected nitrosamines

Nitrosamine	Density (g/mL)	Water solubility (g/100 mL)	Vapor pressure (hPa)*	Boiling point (°C)**
NDMA	1.006 (at 20 °C)	> 10 (at 19 °C)	2.1	151-153
NEMA	0.9448 (at 18 °C)	-	1.1	163
NDEA	0.9431 (at 20 °C)	9.3	0.81	175-177
NDPA	0.9163 (at 20 °C)	0.9894	0.086	206
NDBA	0.8997 (at 20 °C)	slightly soluble	0.03	235
NPIP	1.06 (at 20 °C)	1 – 5 (at 22 °C)	0.092	217-219
NPYR	-	-	0.072	214
NMOR	-	> 10 (at 19 °C)	0.036	224-225

* at 20 °C ** at 760 mmHg
Source: www.chemfinder.com

The International Agency for Research on Cancer (IARC) reports sufficient evidence of carcinogenicity in experimental animals for NDMA and some other nitrosamines (IARC 1974, 1978, 1987, 2000). When administered orally to mice, NDMA induced liver hemangiosarcomas, hepatocellular carcinomas, and kidney and lung tumors. When mice were exposed by inhalation, NDMA induced tumors in lung, liver, and kidney. When administered orally to rats, NDMA induced kidney and bile duct tumors. When rats were exposed by inhalation, it also induced tumors of lung, liver, kidney, and nasal cavity. Also for hamsters, rabbits, guinea pigs, ducks, and fish, tumor induction was observed following exposure to NDMA. Until now, no animal is known in which NDMA does not induce any tumor, supporting the assumption that it is also carcinogenic to humans.

By the U.S. Environmental Protection Agency (EPA) NDMA is classified as "probable human carcinogen" and a 10^{-6} cancer risk level in drinking water at 0.7 ng/L has been determined (EPA 1997). In addition to NDMA, the EPA IRIS database includes eight other aliphatic nitrosamines with significant health concerns which are listed in Table 1.3 with their respective 10^{-6} cancer risk levels. As nitrosamines have not been regarded as drinking water contaminants, U.S. EPA has not yet established a drinking water maximum contaminant level goal (MCLG) or a maximum contaminant level (MCL).

Table 1.3
Cancer risk levels for nitrosamines

Nitrosamine	Short form	10^{-6} cancer risk level (ng/L)
N-Nitrosodimethylamine	NDMA	0.7
N-Nitrosoethylmethylamine	NEMA	2
N-Nitrosodiethylamine	NDEA	0.2
N-Nitrosodi-n-propylamine	NDPA	5
N-Nitrosodi-n-butylamine	NDBA	6
N-Nitrosodiethanolamine	NDELA	10
N-Nitrosopiperidine	NPIP	0.8
N-Nitrosopyrrolidine	NPYR	16
N-Nitrosomorpholine	NMOR	0.8

Source: U.S. Department of Health and Human Services (2005)

California Department of Health Services (DHS) established a notification level in 1998 for NDMA of 10 ng/L. In 2004 and 2005, notification levels of 10 ng/L for NDEA and NDPA were also established. Notification levels are defined by DHS as "health-based advisory levels established by DHS fro chemicals in drinking water that lack maximum contaminant levels (MCLs)" (OEHHA 2006). Response levels, at which DHS recommends removing the source from service, are 100 ng/L for NDEA, 200 ng/L for NDMA, and 500 ng/L for NDPA (DHS 2006). In March 2002, DHS requested a Public Health Goal (PHG) from the Office of Environmental Health Hazard Assessment (OEHHA) and in July 2004 OEHHA initiated a risk assessment for NDMA. In December 2006, OEHHA developed a PHG for NDMA in drinking water of 3 ng/L (OEHHA 2006). This PHG is based on an extra cancer risk of 1 in 1 million for lifetime exposure to NDMA in drinking water.

OCCURRENCE OF NDMA IN DRINKING WATERS

Data on the occurrence of nitrosamines in drinking waters are limited to the United States and Canada. Here, several monitoring programs on the occurrence of NDMA in raw, finished and distributed waters were performed that will be described in this paragraph in more detail.

Concern about NDMA occurrence in drinking water first occurred in Canada in 1986 where treated water from the Oshweken water supply was contaminated with NDMA at levels between 5 and 115 ng/L (Jobb et al. 1994). This discovery prompted an extensive drinking water surveillance program of 145 drinking water plants in Ontario, Canada. The results indicated that NDMA concentrations in the treated water from most of the plants were below 5 ng/L with some samples exceeding 9 ng/L. NDMA was believed to be formed as a result of the reaction of chlorine with treatment chemicals (e.g. amine-based polymers) (Najm and Trussell 2001).

A similar survey was conducted in California in 1999 and 2000, initiated by findings of NDMA in drinking water sources at an aerospace facility in Northern California (ca. 150 ng/L) and in three drinking water wells in Southern California (70-3000 ng/L). Samples from 19 water agencies in the San Francisco bay and Los Angeles area were analyzed for NDMA (DHS 2002; Yamamoto and Sakaji 2003). During this study it was found that only in a few cases NDMA was already present in the influent waters of the drinking water treatment plant. In some of the treatment plants under investigation, however, NDMA was detected in the finished water at levels between 1 and 30 ng/L. Three of 20 chloraminated water samples contained NDMA levels above 10 ng/L; none of the eight treatment plants using free chlorine disinfection had NDMA levels above 5 ng/L; and one of four water supplies using anion exchange treatment had NDMA levels above 10 ng/L. Based on these results it was concluded that chloramination, cationic polymers, and detention time might be key factors for NDMA formation during treatment (Yamamoto and Sakaji 2003, Mitch et al. 2003a).

In 2001 and 2002, a further survey on NDMA occurrence in 21 North American drinking water treatment plants was performed that included treatment plant influent, finished waters (treatment plant effluent) and distribution system waters (Barrett et al. 2003). NDMA was detected in only one out of 81 treatment plant influent water samples. This positive finding was close to the detection limit of 1 ng/L and could not be confirmed during a second sampling campaign. Concentration levels of NDMA in the effluent water samples ranged from below the limit of detection to 30 ng/L, with a median of 1 ng/L for either chlorinated or chloraminated systems. NDMA levels tended to increase in the distribution system, with a median of less than 2 ng/L for chloraminated waters and less than 1 ng/L for chlorinated waters, indicating that this effect was more pronounced for waters disinfected by chloramine. The highest NDMA concentrations were found in a groundwater treated with anion-exchange resin and chlorination. Most samples had NDMA levels below 10 ng/L. More samples with NDMA values between 2.5 and 10 ng/L were found in chloraminated systems than in chlorinated systems, both in treatment plant effluents and distribution systems.

Luo et al. (2003) reported on NDMA concentrations in water samples from 11 different drinking water utilities in mid Missouri, with 10 utilities performing chloramination and one disinfection with free chlorine. NDMA levels in these waters ranged from 3 ng/L up to 48 ng/L. The one from the utility practicing free chlorine disinfection showed a NDMA concentration of 6.1 ng/L. The drinking water from four utilities using chloramination was contaminated with NDMA levels higher than 10 ng/L.

More recently, some studies suggest that water utilities that use effluent dominated surface water sources are prone to NDMA formation during chloramination (Charrois et al. 2004; Pehlivanoglu-Mantas and Sedlak 2006b). At the Oshweken and Brantfort drinking water treatment plants in Ontario (Canada), relying on water from the Grand River with a wastewater portion of about 10%, for example, an average of 13 ng/L NDMA was formed upon chloramination (Pehlivanoglu-Mantas and Sedlak 2006b). Charrois et al. (2004) were able to detect 2-180 ng/L NDMA, 2-4 ng/L NPYR, and about 1 ng/L NMOR in chloraminated drinking water from Alberta (Canada) that was produced from surface water. The authors also observed an increase of NDMA concentrations in a distribution system compared to the finished drinking water effluent of the treatment plant (180 ng/L vs. 67 ng/L).

Some of the NDMA findings in drinking waters might also be linked to a contamination of the raw groundwater or surface water with NDMA. For example, Hoppe et al. (2006) reported on a contamination of the South Platte River by Brighton (Colorado) of up to 6.1 ng/L NDMA. In the wastewater-impacted Quinnipiac River (Connecticut), Schreiber and Mitch (2006a) detected NDMA and NMOR at concentrations of more than 15 ng/L and 100 ng/L, respectively. Contamination of surface water or groundwater by NDMA is most probably caused by the discharge of municipal or industrial wastewater effluents (Mitch et al. 2003a; Quesnel and Nakhla 2005; Schreiber and Mitch 2006a). According to more recent studies, NDMA can be formed at concentrations of up to approximately 1000 ng/L when secondary municipal wastewater is disinfected with chloramine or chlorine prior to surface water discharge or wastewater irrigation (Najm and Trussell 2001; Mitch et al. 2003a, 2005; Siddiqui and Atasi 2004; DHS 2005; Sedlak et al. 2005; Gan et al. 2006; Pehlivanoglu-Mantas et al. 2006; Schreiber and Mitch 2006a).

The studies described above provide first indications concerning possible key factors promoting the formation of nitrosamines during drinking water production, i.e. chloramination or chlorination, long contact times for disinfection, presence of cationic polymers, and surface water sources impacted by waste waters.

RESEARCH APPROACH

There is clear evidence that NDMA is formed during disinfection of waters with chloramine. With regard to the toxicological properties of NDMA (10^{-6} cancer risk level in drinking water is set at 0.7 ng/L by EPA) there is a strong need for strategies and methods to minimize NDMA levels in drinking waters. Furthermore, it is necessary to extend the research to other nitrosamines, as other nitrogen-containing compounds are present in natural waters that undergo disinfection and might react with chlorine or monochloramine by forming other nitrosamines than NDMA.

Thus the main objective of this research project is the development of strategies for minimizing nitrosamine levels in drinking waters. The project does not only focus on formation of the most prominent nitrosamine NDMA but also covers other aliphatic nitrosamines with significant health concern as NEMA, NDEA, NDPA, NDBA, NPIP, NPYR, and NMOR.

For this purpose different approaches were followed. First, information on mechanisms of nitrosamine formation was collected in order to identify options for optimizing operating conditions for disinfection with regard to nitrosamine formation but without compromising disinfection efficiency. Furthermore, significant efforts to identify or at least characterize possible precursors of nitrosamines were undertaken and options for removal of precursor compounds prior to disinfection were studied. Finally, possibilities for nitrosamine removal were evaluated. For all of these approaches (optimization of operating conditions, removal of precursors, removal of nitrosamines), an extensive literature survey was undertaken. Additional laboratory-scale experiments were performed in order to fill knowledge gaps identified during the literature search. The conclusions from the literature search and the laboratory-scale experiments were then verified by full-scale tests in water utilities in U.S., Australia and Germany.

Taking into account the outcome of the literature search as well as the results of the laboratory-based experiments tests with water samples from water utilities, options and strategies for minimizing nitrosamine formation during disinfection of drinking water were derived.

CHAPTER 2
MATERIALS AND METHODS

MATERIALS

All chemicals used for chemical analysis were of analytical grade and commercially purchased. Nitrosamines and deuterated nitrosamines had a purity ≥ 99% and were obtained from Dr. Ehrenstorfer GmbH (Augsburg, Germany). Aliphatic amines (purity ≥ 98%) and 9-fluorenylmethyl chloroformate (FMOC-Cl) were from Sigma-Aldrich Chemie GmbH (Taufkirchen, Germany).

For the NFP tests, ammonium chloride (Trace Select, > 99.5% purity) from Fluka (Buchs, Switzerland), caustic soda (10%, zur Analyse) from VWR International GmbH (Darmstadt, Germany), and a sodium hypochlorite solution (6-14% as free chlorine) from Riedel-de Haën (Seelze, Germany) were used. Buffer solutions were prepared from potassium dihydrogenphosphate (KH_2PO_4, zur Analyse) and disodium hydrogenphosphate (Na_2HPO_4•2 H_2O, zur Analyse), both purchased from VWR International GmbH. For quenching the chloramine and chlorine solutions, sodium sulfite (zur Analyse) from VWR International GmbH was used. All chemicals tested in the NFP test were also obtained from Sigma-Aldrich Chemie GmbH. They were of analytical grade and were used without further purification.

For the preparation of stock solutions, MilliQ water prepared in a Millipore system (Millipore GmbH, Schwalbach, Germany) was used. A 24 mM aqueous ammonium chloride solution was prepared by dissolving 1.3 g ammonium chloride in 1 L of MilliQ water and adjusting the pH to 8.0 by addition of small amounts of caustic soda. For the NFP experiments, a chloramine stock solution was prepared by adding hypochlorite solution to this aqueous ammonium chloride solution. Prior to the preparation of the chloramine solution, the free chlorine concentration in the hypochlorite stock solution was determined by a DPD test kit LCK 310 from Dr. Lange (Hach-Lange GmbH, Düsseldorf, Germany). Based on the free chlorine concentration in the hypochlorite solution, the volume of hypochlorite stock solution was calculated that was necessary to achieve a molar ratio ammonia:free chlorine of 1.2:1. The respective volume of hypochlorite stock solution was then added to 500 mL of ammonium chloride solution. Before and after each NFP test, the concentration of free and bound chlorine was determined in each monochloramine solution by the DPD test kit LCK 310 from Dr. Lange.

Iron chloride ($FeCl_3$) and polyaluminum chloride (Sachtoklar 27S) for flocculation experiments were purchased from Giulini Chemie (Ludwigshafen, Germany) and Sachtleben Chemie GmbH (Duisburg, Germany), respectively.

Ozone stock solutions for the ozonation experiments were produced by sparging O_3-containing oxygen through Milli-Q water that was cooled in an ice bath. Ozone concentration in the stock solutions was ca. 1.5 mM. Exact concentration levels were determined by direct UV measurement at 258 nm ($\varepsilon(\text{ozone})_{258\ \text{nm}} = 3000\ M^{-1}cm^{-1}$).

Chlorine dioxide was produced according to a method described by Hoigné and Bader (1994).

Activated carbon F300 for adsorption experiments was from Chemviron Carbon (Feluy, Belgium) and was used in powdered form (average particle size ca. 65 µm).

TRACE-LEVEL ANALYSIS OF NITROSAMINES IN WATER SAMPLES

The methodology used for the simultaneous determination of several carcinogenic N-nitrosamines in aqueous samples at very low ng/L levels was newly developed and is based on a solid-phase extraction procedure followed by gas chromatography with large-volume injection (LVI) technique coupled to a quadrupole mass spectrometer with positive chemical ionization (PCI) using isobutene as reagent gas. Target analytes include NDMA, NEMA, NDEA, NDPA, NDBA, NMOR, NPYR and NPIP.

Sample preparation is based on a mixed bed solid-phase extraction (SPE) procedure in which the SPE cartridge consists of two layers: an activated carbon (AC) layer at the bottom and a non-polar C18 sorbent layer at the top of the cartridge. This design guarantees that active AC sites, necessary for the retention of nitrosamines, are not blocked by non-polar matrix constituents, since such compounds are retained in the upper C18 layer. Mixed bed SPE cartridges (6 mL) were self-packed with 1.5 g activated carbon (for GC, 35-50 mesh ASTM, Merck, Darmstadt, Germany) as bottom layer and 0.5 g Isolute C18(EC) sorbent (International Sorbent Technology Ltd., Mid Glamorgan, UK) as top layer. Between both sorbents as well as at the bottom and the top of the cartridge a polyethylene frit was placed. SPE columns were conditioned with 20 mL dichloromethane, followed by 10 mL methanol and 10 mL water (containing 0.8 g/L sodium sulfite). 500 mL water samples in amber bottles (containing 0.8 g/L sodium sulfite) were adjusted to pH 3 with hydrochloric acid and spiked with 40 μL of an internal standard solution containing NDMA-d6, NDEA-d10, NDPA-d14, and NPYR-d8 in acetone (final concentration in the sample 40 ng/L). Water samples were loaded onto the SPE columns automatically using an AutoTrace SPE Workstation from Caliper Life Sciences GmbH (Rüsselsheim, Germany) at 3 mL/min. Following sample loading the cartridge was dried under positive nitrogen pressure for 4 hours. Analytes were eluted manually with 12 mL dichloromethane and the eluate was dried over sodium sulfate. Finally, the eluate was gently concentrated to a final volume of approximately 150 μL using a rotary evaporator (HB-140 Rotavapor M, Büchi, Switzerland). The residual sample extracts were then pipetted into micro vials and transferred to the GC vials for analysis.

Nitrosamines were chromatographically separated using an Agilent (Waldbronn, Germany) 6890N gas chromatograph (GC) coupled with an Agilent 5973 inert mass selective detector (MSD), operated in positive chemical ionization (PCI) mode. The capillary column was a DB-WAX (30 m x 0.25 mm ID x 0.5 μm film thickness, J&W, supplied by Agilent). Helium was used as the carrier gas at an inlet pressure of 8.9 psi, resulting in a column flow of 1.2 mL/min (constant flow mode). PCI reagent gas was isobutane (3.5, Air Liquide GmbH, Ludwigshafen, Germany) with the flow control set at 1.3 mL/min (26%). 50 μL aliquots of the extracts were injected using an MPS2 automatic liquid sampler in combination with a CIS 4Plus cooled injection system with Peltier cooling (Gerstel, Mülheim/Ruhr, Germany) via the solvent vent mode. The inlet of the cooled injection system was equipped with a deactivated single necked glass liner filled with glass wool (2 mm inner diameter, 3 mm outer diameter, 71 mm length, pesticide grade, Gerstel). The GC oven temperature was held at 35 °C for 3 min, then heated at 10 °C/min to 220 °C which was held for 8.5 min (run time: 30 min). The transfer line temperature was kept at 250 °C. The MSD source and quadrupole temperatures were set at 250 °C and 150 °C, respectively. The mass spectrometer was operated in the PCI mode with a solvent delay of 9 min. Nitrosamines were each quantified by selected ion monitoring (SIM mode) of the corresponding protonated molecule ion $[M+H]^+$.

The method showed a linear relation between peak area and concentration in the investigated concentration range of 0 to 100 ng/L and was free of any interference (< 0.2 ng/L). This was also controlled by analyzing samples spiked with the parent amines only (e.g., 10 μg/L dimethylamine). Additionally, peak verification was performed for all samples by measuring the same sample extract twice, once prior and once after its irradiation with UV light. Overall extraction efficiencies from spiked water samples were independent of the investigated matrix and amounted to 50-70%. Determination of the limits of quantitation (LOQs) was performed according to the German standard procedure DIN 32645 from the linear regression analysis based on calibration samples that were analyzed via the complete procedure (incl. sample preparation) (DIN 1994). Determined LOQs were about 1 ng/L for all nitrosamines. Using spiked drinking water and surface water samples with known nitrosamine concentrations, coefficients of variation for within-day repeatability and between-day precision were in general less than 20% at the LOQs and less than 9% at higher concentration levels.

Water samples, standards and sample extracts were protected from open sunlight at all times to prevent photodegradation. All glassware, which was used during sample preparation several times, was intensively rinsed manually with purified water and acetone and was then pyrolyzed at 450-500 °C for 2 hours. When stored at 4 °C in the dark, aqueous nitrosamine samples were shown to be stable for at least 4 weeks.

TRACE-LEVEL ANALYSIS OF ALIPHATIC AMINES IN WATER SAMPLES

Analysis of amines was performed by a newly developed approach based on solid-phase derivatization with 9-fluorenylmethyl chloroformate (FMOC-Cl) and subsequent analysis via high-performance liquid chromatography (HPLC) coupled with fluorescence detection. Using this method, primary and secondary aliphatic amines can be determined whereas tertiary amines are not captured (Pietsch et al. 1996).

The solid-phase derivatization procedure is based on a simultaneous analyte enrichment and derivatization on a solid-phase extraction (SPE) cartridge. For analysis, a 500 mL volume of water sample, being stabilized with 400 mg/L sodium azide, was spiked with both 1 g EDTA tetrasodium salt and 25 μL of an internal standard solution containing n-hexylamine and n-heptylamine in acetonitrile (final concentration in the sample: 2500 ng/L). Finally, the sample was adjusted to pH 11.8 with sodium hydroxide. For solid-phase derivatization a Lichrolut EN (40-120 μm, 200 mg, 6 mL, Merck, Darmstadt, Germany) SPE cartridge was used. The SPE material was conditioned with 5 mL of an acetonitrile/ethylacetate mixture (1:1) followed by 5 mL pure acetonitrile and 500 μL of a 12 g/L FMOC-Cl solution in acetonitrile. Then, the sample was applied to the SPE cartridge with a vacuum device (VacMaster, IST). Following solid-phase derivatization, the SPE cartridge was dried under a gentle steam of nitrogen for one hour, and the derivatized analytes were eluted with four times 2.5 mL of an acetonitrile/ethylacetate mixture (1:1). The eluate was then evaporated in a sample concentrator (DR 3A, Techne) to complete dryness (with nitrogen), re-dissolved in 1.0 mL acetonitrile, and finally transferred to appropriate sample vials for HPLC analysis.

The HPLC system (Agilent 1100-Series) used for amine analysis consisted of a G1311A quaternary pump, a G1321A fluorescence detector, a G1316A thermostatted column compartment, and a G1313A autosampler. For on-line degassing of the mobile phases, a G1322A degasser was utilized. The analytical column was a Gemini C18 110A (3.0 x 250 mm, 5 μm particle size) obtained from Phenomenex (Aschaffenburg, Germany). The column was protected by a pre-column Gemini C18 (2.0 x 4 mm, Phenomenex). Evaluation of chromatograms was performed by fluorescence detection using wavelengths for excitation of 280 nm and for emission of 305 nm. The oven temperature was kept at 40 °C. The injection volume was 5 µL. A multilinear binary gradient was formed from eluent A, water, and eluent B, pure acetonitrile. The flow rate was adjusted to 0.6 mL/min and the gradient settings were as follows: 0 min, 45% B; 5 min, 45% B; 45 min, 70% B; 55 min, 90% B; 60 min, 100% B; 65 min, 100% B, and 70 min, 45% B.

To compensate for analyte losses during sample preparation, n-heptylamine and n-hexylamine were used as internal standards. Both compounds were found to be absent from environmental samples. The method showed a linear relation between peak area and concentration in a concentration range of 0 to 200 ng/L. For calibration, samples running through the whole procedure were used. Limits of detection (LOD) and limits of quantitation (LOQ) were determined according to DIN 32645 from the linear regression analysis and revealed LODs in the range of 10 to 30 ng/L, and LOQs of 30 to 100 ng/L, depending on the compound of interest. Using spiked drinking water and surface water samples with known amine concentrations, coefficients of variation for within-day repeatability and between-day precision were in general less than 22% at the LOQs and less than 12% at higher concentration levels. When stabilized with sodium azide and stored at 4 °C in the dark, samples were shown to be stable for at least 2 weeks.

DETERMINATION OF THE NITROSAMINE FORMATION POTENTIAL (NFP)

NFP tests were performed to characterize water samples with regard to their potential to form nitrosamines. The levels of nitrosamines formed in this test can be used to compare the potential of different waters to form nitrosamines during chloramination as the amount of nitrosamines formed characterizes the quantity and potency of precursors present in the sample under investigation. However, the NFP test is not thought to predict concentration levels of nitrosamines formed during disinfection in waterworks or in distribution systems.

The experimental design of the NFP tests follows closely the procedure described as nitrosamine precursor test by Mitch et al. (2003b). Unless stated otherwise, the NFP tests were performed at a pH of 7 which was achieved by adding 700 mg potassium dihydrogenphosphate and 880 mg disodium hydrogenphosphate to 980 mL of the water sample. Then, 20 mL of the chloramine stock solution was added to the water sample, resulting in a chloramine dose of 0.4 mM (= 30 mg/L as Cl_2). Chloramine concentration of the stock solution as well as initial and final chloramine concentration in the waters under investigation was determined using Dr. Lange test kit LCK 310/343 which is based on a colorimetric detection of a red dye formed by a reaction of diethyl-p-phenylenediamine with the oxidizing agent. In all experiments, chloramine concentrations did significantly decrease with reaction time mainly due to its self-decomposition but the final concentration levels were independent of the water type under investigation and thus the chloramine exposure should be comparable in all experiments. The same effect was observed by Mitch and Sedlak (2002a) who investigated the NDMA formation potential of effluents from different waste water treatment plants.

Prior to the NFP test, pH adjustment of the samples was done by addition of phosphate buffers. All experiments were performed in 1 L brown glass flasks which were stored at 20 °C in the dark. After 7 days, the residual chloramine concentration was measured and 2.5 g/L sodium sulfite were added in order to prevent further nitrosamine formation. Then, the samples were analyzed for nitrosamines. Since some of the tested waters already contained low background levels of nitrosamines (1 to 5 ng/L) before chloramination, each original water sample was analyzed for these background levels in order to correct the results of the NFP test.

The impact of the experimental conditions on the results of the NFP test was studied in detail during the validation procedure which is described in Chapter 3.

LABORATORY-SCALE EXPERIMENTS FOR SIMULATION OF TREATMENT STEPS

Biodegradation

The biodegradation of nitrosamines as well as possible precursor compounds was studied using an established test filter device for simulation of the aerobic microbial degradation during riverbank filtration. In recent years, this experimental method has been adapted in order to study the behavior of single substances during the riverbank filtration process at concentration levels relevant to the environment (Karrenbrock et al. 1999; Knepper et al. 1999). The test filter devices have been setup according to the scheme presented in Figure 2.1. Each test filter consists of (i) a filter column filled with carrier material, (ii) a reservoir bottle (15 L conical shoulder bottle) with air supply and sampling device, and (iii) a circulation system with a controllable diaphragm pump (CFG Prominent, Heidelberg, Germany). As material for tubes, tube couplings, and sampling cocks PTFE was chosen.

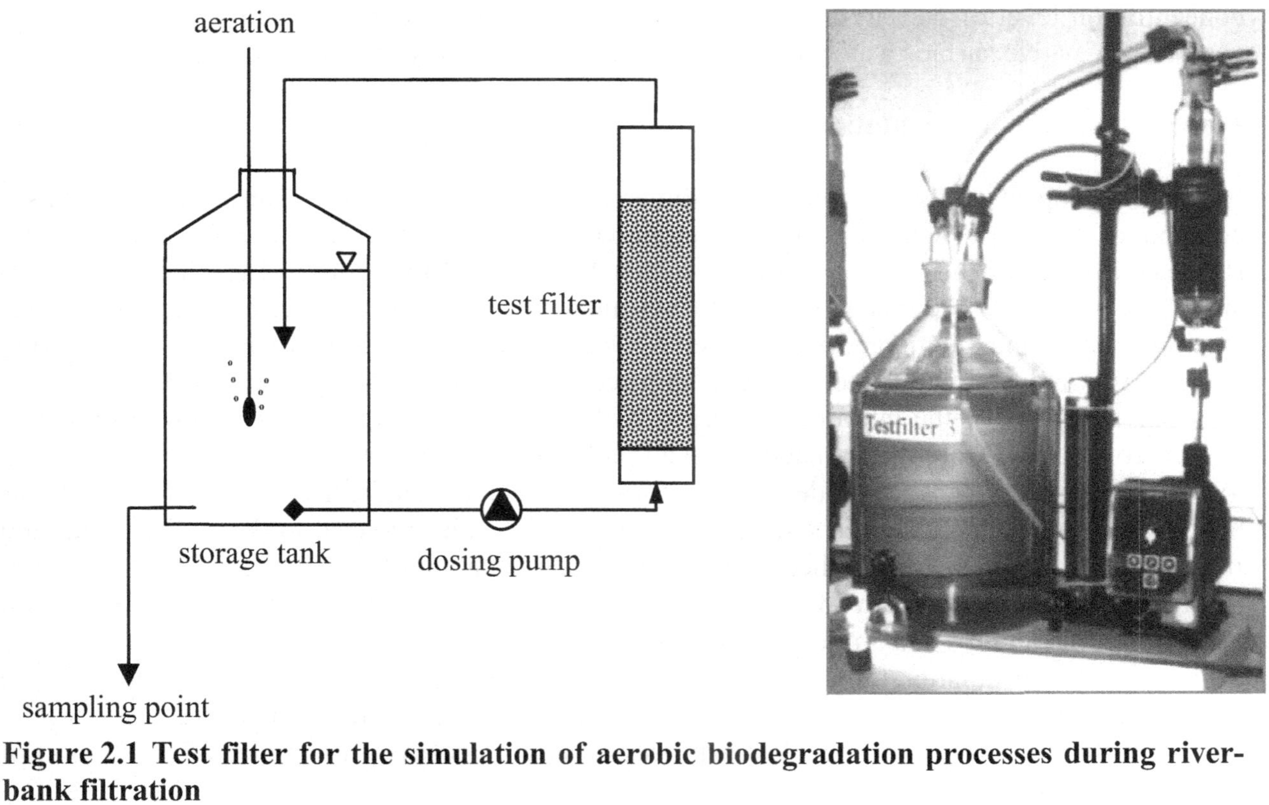

Figure 2.1 Test filter for the simulation of aerobic biodegradation processes during riverbank filtration

Each test filter column (diameter: 5 cm; height: 23 cm) has an empty volume of 465 mL. The carrier material consists of porous sintered glass beads with a defined pore size distribution (SIRAN-Carrier No. 023/02/300, Schott Engineering GmbH, Mainz, Germany) and was fixed within the column by an apertured glass bottom. The material typically provides no adsorption capability and is normally used for immobilization of microorganisms in biotechnological applications. Furthermore, it is chemically resistant, easy to clean, and can be sterilized by high-temperature steaming. The sintered glass beads have a diameter of 2-3 mm and a surface area of 80 m^2/L or 0.2 m^2/g. The material provides a high pore volume of 55-60% with a pore diameter of 60 to 300 µm. Filter bed height is 18 cm and bulk density roughly 450 g/L.

The test is executed by filling the reservoir with 10 L of non-filtered surface water and spiking it with the test compound under investigation at concentrations that are relevant to the environment. For a time period of about 30 days, the water is then pumped under re-circulation, at 20 °C, and in the dark over the carrier material in the filter column. The flow rate is adjusted to a value between 10 and 15 mL/min. The porous material provides optimal conditions for microbial colonization and in the course of time, diverse microorganisms are established and are growing unhindered on the glass beads, being principally capable of degrading the test substance, provided that it is biodegradable. The system is permanently aerated with ambient air to assure aerobic conditions. During the test period, samples are taken after defined periods of time for determination of test substance concentrations. From the time-dependence of the concentration levels, the microbial degradation of the compound under investigation can be assessed.

The studies concerning the microbial degradability of nitrosamines were performed at a concentration level of 100 ng/L. For investigation of the removability of nitrosamine precursor compounds, native surface water samples were taken without any spiking.

Flocculation and Sedimentation

The impact of flocculation and sedimentation on the removal of nitrosamine precursor compounds was investigated in a non-continuous laboratory-scale test, the so-called Jar test (DVGW 1998). For this test, a Stuart flocculator SW6 from Bibby Sterlin Ltd., Stone Staffordshire, UK was used. The system consists of six units with 1 L glass beakers and stirrers with defined geometry and submerged stirring depth. During the flocculation experiments, the stirring velocity was reduced with time, resulting in a reduction of the velocity gradient (G value) and thus in a reduction of the introduced energy. With decreasing energy, the formation of colloids and micro-flocs is favored. When adding an aliquot of the stock solution of the flocculation agent to 800 mL of the river water under investigation, the velocity gradient G was at 300 s^{-1}. It was held at this value for 20 s, then reduced to 70 s^{-1} for 2.5 min, reduced to 40 s^{-1} for 5 min and finally reduced to 25 s^{-1} for another 10 min. After sedimentation of the flocs for 30 min, 500 mL of the supernatant were taken for performing the NFP tests. Before and after each flocculation experiment, pH and UV adsorption at 254 nm were measured.

For the flocculation tests, polyaluminum chloride ($pAlCl_3$) and iron chloride ($FeCl_3$) were tested as flocculants and were added at doses of 10 mg/L Al, and 5 and 10 mg/L Fe, respectively. A pH regulation during the flocculation process was not necessary as immediate formation of flocs was observed in all experiments.

Oxidation

The second order rate constants for reactions of oxidants with nitrosamines and possible precursor compounds were determined using either of two different kinetic methods. The first one is measuring the decrease of oxidant (or compound) in an excess of compound (or oxidant) to fulfill pseudo first-order conditions. The second one is applying the concept of competition kinetics with a reference compound. Details of this kinetic methods are available in Huber et al. (2003).

The rate constants for reactions of ozone with NDMA, dimethylethanolamine (DMEA), and Dimethylformamide (DMFA) were determined by measuring the decrease of ozone in the presence of an excess of compound. The reaction was initiated by injecting ozone of 20-40 μM to buffered solution of 1 mM-0.1 M compound at pH 2.5-4, depending on the rate constants of the reactions. The rate constants for reactions of ozone with dimethylaminobenzene (DMAB) and 4-dimethylaminoantipyrine (DMAP) were determined by measuring the decrease of compound in the presence of an excess of ozone. The reaction was initiated by injecting 20 μM ozone to buffered solution of 1-2 μM compound at pH 1-2.5. Samples were withdrawn at predetermined time intervals and the concentration of ozone or the compound was analyzed immediately after quenching in an indigo solution. The rate constants of DMDC and DMAI were determined by competition kinetics using phenol ($k_{O3,\ phenolate} = 1.4 \times 10^9\ M^{-1}s^{-1}$; Hoigné and Bader 1983) as a reference compound at pH 7-10.

The rate constants for reactions of chlorine dioxide with NDMA, DMEA, and DMFA were determined by measuring the decrease of chlorine dioxide in the presence of an excess of compound (10 mM-0.1 M) at pH 3.5-7, depending on the rate constants of the reactions. The rate constants of DMDC and DMAI were determined by measuring the decrease of compound concentration (1-2 μM) in the presence of an excess of chlorine dioxide (50 μM) at pH 4-6, and 10, respectively. Competition kinetics using phenol ($k_{ClO2,\ phenolate} = 2.7 \times 10^7\ M^{-1}s^{-1}$; Neta et al. 1988) as a reference compound was used to determine the rate constant of DMAP at pH 7 to 9.

The rate constants for reactions of hydroxyl radical with NDMA, DMA, DMEA, DMTC (dimethyldithiocarbamate), DMAI (3-(dimethylaminomethyl)indole), and DMAP were determined by competition kinetics (Haag and Yao 1992), using p-chlorobenzoic acid as reference compound ($k_{pCBA,\bullet OH} = 5 \times 10^9\ M^{-1}s^{-1}$; Buxton and Greenstock 1988). $^{\bullet}OH$ was produced by an O_3/H_2O_2 system for NDMA, by a UV/H_2O_2 system for DMA and DMEA, and by γ-radiolysis for DMDC, DMAI and DMAP.

The pH-dependent rate constants of several compounds in higher pH regions were extrapolated from those obtained in lower pH condition. The rate constants for the reactions not stated above were taken from the literature (Hoigné and Bader 1983; Huie and Neta 1986; Buxton and Greenstock 1988; Hoigné and Bader 1994).

Activated Carbon Filtration

For studying the removal of precursor compounds for nitrosamine formation by adsorption onto activated carbon, batch experiments with different amounts of powdered activated carbon (PAC) were performed. For each set of experiments, different amounts of a F300 type PAC (1, 5, 10, 20, 50, and 100 mg/L) were added in a 1 L brown glass bottle to 600 mL of river water. All glass bottles were shaken for 7 days at room temperature to assure that adsorption equilibrium is achieved as nothing is known about adsorption kinetics of the precursor compounds yet.

Then, the PAC was removed by a folded filter (size 595½ from Schleicher & Schuell, Dassel, Germany). In parallel to the samples which were in contact with the PAC, two samples without PAC contact were investigated: One which has been treated as the others (i.e. shaken for ten days and filtered as the others) but without addition of PAC and one which has only been stored at 4 °C. As long as both samples revealed the same NFP, there is no evidence for losses of precursor compounds due to microbial degradation or adsorption onto glass surfaces or filter materials.

Based on the results of these adsorption experiments, a first assessment of the adsorbability of the precursors onto activated carbon can be based on a calculation of adsorption isotherms. Using this approach, equilibrium partition of the precursor compounds between the aqueous phase and the activated carbon is assumed and no kinetic effects are taken into account. From the NFP determined in the filtered water samples, the evaluation of the isotherms was performed according to the Freundlich equation $q = K\,c^{1/n}$, where q denotes the capacity of the coal, c the concentration of the precursors in the water sample, and K and 1/n the Freundlich parameters (Freundlich 1906; Sontheimer et al. 1988). 1/n mainly describes the concentration dependence of the adsorbability whereas K characterizes the adsorption capacity of the charcoal for the compounds under investigation. As a rule, the higher the K value the better the adsorbability of a compound. According to the FREUNDLICH equation a straight line is obtained for the adsorption of a single compound if q is plotted as a function of c in double-logarithmic scale. For mixtures of compounds, however, the isotherm may deviate considerably from a straight line, depending on the differences in adsorption characteristics between the compounds in the mixture (Sontheimer et al. 1988).

CHAPTER 3
FORMATION OF NITROSAMINES

LITERATURE REVIEW

The chemistry of nitrosamine formation in general is very complex. Many reactions can occur simultaneously and the products depend on the exact reaction conditions, including concentrations of reactants, catalysts and inhibitors, and the relevance of other reactions which can compete for the nitrosating agents. Furthermore, most of the published work focuses on NDMA formation, while formation of other nitrosamines has attracted only little attention in the past. However, most of the principles directing NDMA formation should, at least on a qualitative basis, also be valid for the formation of other nitrosamines.

According to the literature studied, there are three major pathways by which nitrosamines might be formed:

- nitrosation of nitrogen-containing compounds by nitrosating agents (mainly nitrous acid),
- reaction of inorganic chloramines as monochloramine with aliphatic amines (e.g. dimethylamine) to unsymmetrical dimethylhydrazine (UDMH) and subsequent oxidation to nitrosamines and
- reaction of dichloramine with aliphatic amines to chlorinated UDMH which can be further oxidized to nitrosamines by either monochloramine or dissolved oxygen.

Nitrosamine Formation by Nitrosation of Nitrogen-Containing Compounds

Nitrosation means the introduction of a nitroso group (-NO) into an organic compound causing the formation of nitroso compounds. By N-nitrosation of nitrogen-containing organic compounds nitrosamines are formed. While primary nitrosamines suffer from limited stability and may react further to alcohols and olefins, secondary and tertiary nitrosamines are quite stable (Cohen and Bachman 1978).

Possible nitrosating agents are summarized in Table 3.1 and include nitrous acid (HNO_2), some nitrogen oxides (N_2O_3, N_2O_4), nitrosyl halides and thiocyanates (NOCl, NOBr and NCSNO) (Mirvish 1975; Wainright 1986).

Table 3.1
Nitrosating agents and precursors for nitrosamines

NO^+	nitrosonium ion
$H_2NO_2^+$	nitrous acidium ion
HNO_2	nitrous acid
N_2O_3	nitrous anhydride
N_2O_4	dinitrogen tetroxide
XNO (X = Cl, Br)	nitrosyl halide
NCSNO	nitrosyl thiocyanate
C_6H_5NO	nitrosophenol

Source: Wainright (1986)

However, not all nitrogen oxides are able to cause nitrosation. Nitrous oxide (N_2O) and nitric oxide (NO) for instance do not form nitrosamines. Nitrates do not form nitrosamines directly, but might be transformed to nitrites either by microorganisms present in the environment or by naturally occurring bacteria in the mouth (Preussmann 1984; Vermeer et al. 1998).

Other possible nitrogen containing precursors for N-nitrosamines besides the aliphatic and aromatic amines are N,N-dialkylhydrazides, N-alkylamides, N-alkylureas, 1,3-dialkylureas, and N-alkylcarbamates for the monoalkylnitrosamines, and 1,1-dialkylureas, 1,3-dialkylphenylureas, tetraalkylureas, and 1,1-dialkylthioureas for the dialkylnitrosamines (Mirvish 1975; Fishbein 1979). Of course, also organic compounds having structural elements of the above mentioned compounds can act as precursors for nitrosamine formation, including various pesticides, cosmetics, sunscreen agents, pharmaceuticals (Loeppky et al. 1994) or even natural organic matter (NOM). A more detailed overview on nitrosamine precursors is given in Chapter 4.

Also tertiary amines are able to form nitrosamines (Ayanaba and Alexander 1974) but are mostly regarded as less reactive towards nitrosating agents. Mirvish (1975), for instance, studied the formation of NDMA and NDEA from the aliphatic amines trimethylamine and triethylamine and found that nitrosamine formation from trialkylamines was too slow to give a significant contribution and thus was not regarded as important. In the review of Wainright (1986), however, it is mentioned that at pH 6 NDMA is formed more rapidly from trimethylamine than from dimethylamine. In addition to this, Loeppky et al. (1994) found that some tertiary N,N-dialkyl aromatic amines quite quickly generate nitrosamines.

Quaternary amines having four organic substituents on the nitrogen (i.e. not being protonated tertiary amines) may also form nitrosamines (Wainright 1986) and so do amine oxides like trimethylamine-N-oxide (Ohshima and Kawabata 1978). In general, however, the yields of nitrosamine formation from quaternary amines are rather low as the initial reaction step necessitates dealkylation which is not supported by nitrosating agents. Some papers also mention the possibility of transnitrosation, e.g. from aromatic nitrosamines to aliphatic amines, as possible source of aliphatic nitrosamines (Fishbein 1979; Singer et al. 1978).

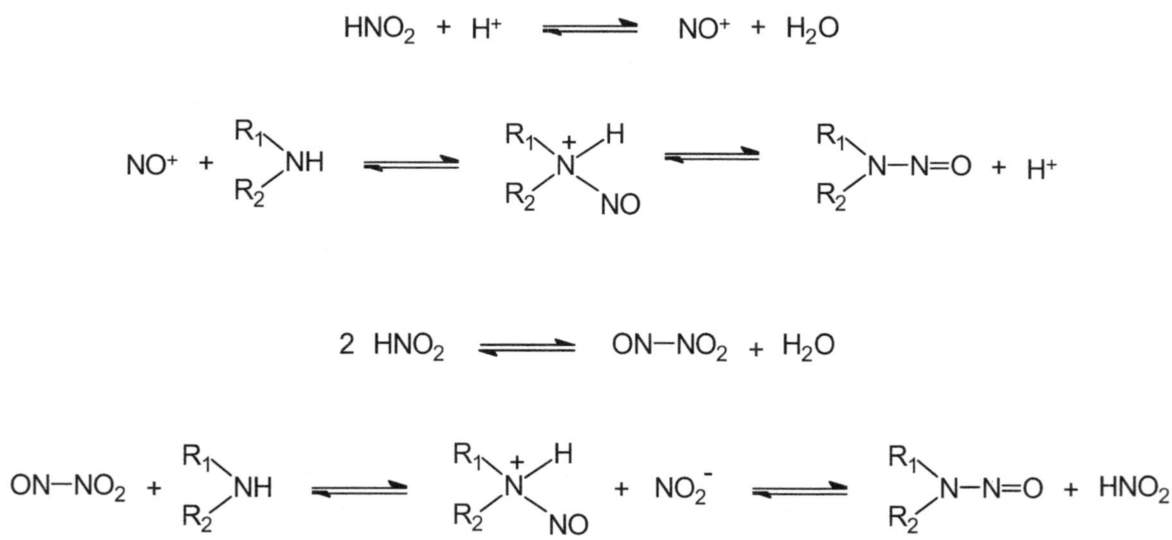

R_1, R_2: Alkyl groups, e.g. methyl or ethyl

Figure 3.1 Formation of nitrosamines by nitrosation of secondary amines

The general principle of the nitrosation reaction induced by nitrite is as follows: Under acidic conditions nitrite (NO_2^-) is transformed to nitrous acid (HNO_2, $pK_a = 3.37$) which is not stable in aqueous solution but decomposes either to the nitrosyl cation NO^+ or to dinitrogen trioxide N_2O_3 (Mirvish 1975), both of them being capable of reacting with nitrogen-containing organic compounds to form nitrosamines. Figure 3.1 shows as an example the reaction pathway for the nitrosation of a dialkylamine by NO^+ and N_2O_3.

Nitrosation Reaction Kinetics

Most often, kinetics of nitrosamine formation is determined by the formation of the nitrosating agent (e.g. N_2O_3), as this step is slower than the nitrosation step of the amine (Kalatzis and Ridd 1966). At low pH, however, the availability of the non-protonated amine is limited and, thus, kinetics of nitrosamine formation is determined by the nitrosation of the amine. Mirvish gives rate constants k_1 and k_2 of the nitrosamine formation by the reaction

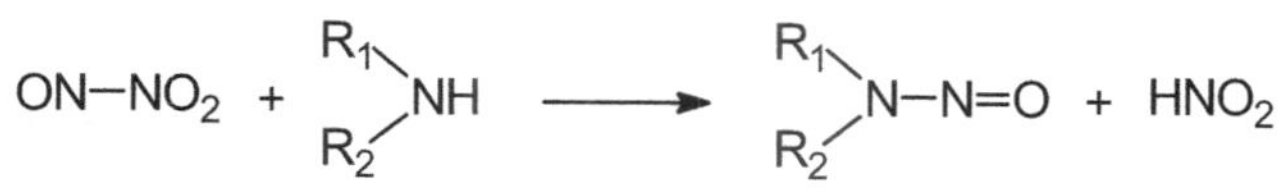

k_1 and k_2 denote rate constants defined by the equations

$$\text{Nitrosamine formation rate} = d(\text{nitrosamine})/dt = k_1\,(R_2NH)\,(HNO_2) \text{ and}$$

$$\text{Nitrosamine formation rate} = d(\text{nitrosamine})/dt = k_2\,(\text{amine})\,(\text{nitrite}).$$

In the first equation (R_2NH) and (HNO_2) represent the concentrations of the non-protonated amine and the free protonated nitrous acid. The concentrations of both species depend on the pH and are difficult to determine. The constant k_1 itself, on the other hand, is independent of pH. In the second equation (amine) and (nitrite) represent the total concentrations of amine and nitrite which do not depend on pH. In this second equation, the pH dependence of nitrosamine formation is implemented in k_2, which consequently depends on the pH of the solution.

The k_1 and k_2 values published by Mirvish (1975) for several secondary and tertiary amines are presented in Table 3.2. The values for k_2, valid for the optimum pH of nitrosamine formation, are also given in Table 3.2 (please note that the optimum pH given in Table 3.2 refers to the formation of nitrosamines from reaction of N_2O_3 with the aliphatic amine and does not take into account N_2O_3 formation).

The similar k_1 values for different amines in Table 3.2 indicate that amines (in the non-ionized form) provide quite similar reactivities towards N_2O_3. Assuming that k_1 is constant, k_2 can be calculated from the pK_a value of the amine using an equation also derived from Mirvish (1975):

$$\log(k_2 \text{ in } M^{-2}s^{-1}) = 7.8 - pK_a.$$

This equation implies that nitrosamine formation is faster for amines of weak basicity than for amines with high basicity (Fishbein 1979) and, thus, amines with weak basicity such as morpholine will nitrosate faster than e.g. dimethylamine (Fan and Tannenbaum 1973; Challis et al. 1978).

Table 3.2
Rate constants k_1 and k_2 for the nitrosation of amines at optimum pH and 25 °C

Amine	Optimum pH	k_1 in 10^5 $M^{-2}s^{-1}$	k_2 in $M^{-2}s^{-1}$
Piperidine	11.2	1.4	0.00045
Dimethylamine	10.72	1.5	0.0017
Pyrrolidine	11.27	21.0	0.0053
N-Methylethanolamine	9.5	0.62	0.010
N-Methylbenzylamine	9.54	0.92	0.013
Proline	-	1.4	0.037
Sarcosine	-	2.6	0.23
Prolylglycine	8.97	5.0	0.25
Hydroxyproline	-	2.1	0.31
Prolylleucylglycineamide	8.97	6.2	0.38
Morpholine	8.7	2.3	0.42
Mononitrosopiperazine	6.8	0.83	6.7
Aminopyrine	5.04	1.0	80
Piperazine	5.57 and 9.8	0.62	83
N-Methylaniline	4.85	18.0	250

Source: Mirvish (1975)

Table 3.3 gives an overview of pK_a values of aliphatic amines and the respective k_2 values calculated according to the equation of Mirvish (1975). It can be seen that, e.g. for morpholine and piperazine, the data calculated from pK_a values are in good accordance to the rate constants in Table 3.2 deduced by Mirvish (1975).

Table 3.3
pK_a values for aliphatic amines and rate constants k_2 calculated according to the equation of Mirvish

Amine	pK_a	k_2 in $M^{-2}s^{-1}$
Methylamine	10.6	0.0016
Dimethylamine	10.7	0.0013
Ethylamine	10.8	0.0010
Diethylamine	11.0	0.00063
n-Propylamine	10.7	0.0013
n-Butylamine	10.8	0.001
n-Pentylamine	10.6	0.0016
n-Hexylamine	10.6	0.0016
n-Heptylamine	10.0	0.0063
n-Octylamine	10.7	0.0013
Cyclohexylamine	10.7	0.0013
Morpholine	8.2	0.398
Pyrrolidine	11.3	0.00032
Piperidine	11.2	0.00040
Piperazine	9.8/5.6	0.010/158
Ethylendiamine	11.8/0.8	0.00010
Hexamethylendiamine	10.7/7.6	0.0013
Ethanolamine	9.5	0.020
Diethanolamine	8.9	0.079

Source: Pietsch (1997); Mirvish (1975)

As it can be seen from the reaction scheme in Figure 3.1, the formation of the nitrosyl cation NO^+ as well as the nitrosation of the aliphatic amine depends on the pH. It was found that nitrosamine formation by nitrosation of dimethylamine is most rapid at pH 3.4 (being the pK_a of nitrous acid) and is rather slow at neutral and basic pH values (Mirvish 1970, 1975; Fan and Tannenbaum 1973). For piperidine, morpholine, and mononitrosopiperazine Mirvish found an optimum at pH 3.0 (Mirvish 1975). For morpholine, also an optimum pH for nitrosation of 3.36 is reported (Fan and Tannenbaum 1973). Other authors reported on an optimum pH for nitrosation of morpholine at 3.0 or 3.36 and for proline at 2.25 (Wainright 1986). Fishbein (1979) and also Mirvish (1977a, 1977b) in his later papers, however, emphasize that, although the initial reaction rate for nitrosation of secondary amines has its maximum at pH 3.0 to 3.4, the yield for prolonged reaction periods is largest at pH 4 to 5. This maximum reaction yield at pH 4 to 5 is explained by the faster decomposition of nitrous acid at lower pH values. Nitrosation is, however, also dependent on the amine not being protonated and decreasing the pH reduces the availability of the reactive amine species. Therefore, there is always a pH optimum for nitrosation which reflects a balance between these opposing conditions.

Mills and Alexander (1976) also found a strong impact of pH on nitrosamine formation when they studied the formation of NDMA from dimethylamine and nitrite in soils, sewage, and lake water. At pH 8.7, they could detect no nitrosamine formation whereas at pH 5.2 significant amounts of NDMA were found when they added dimethylamine and nitrite to sewage water samples. A more effective formation of nitrosamines was also found by Yordy and Alexander (1981) who studied the formation of N-nitrosodiethanolamine (NDELA) from diethanolamine in lake water and sewage. Nitrosamine formation was shown to be also pH-dependent in soils (Pancholy 1978).

Nitrosation might be enhanced via catalysis by metal salts (e.g. Cu, Fe, Zn) (Challis et al. 1978), by nucleophilic anions (e.g. thiocyanate and to a smaller extend halides) (Boyland et al. 1971; Fan and Tannenbaum 1973; Mirvish 1975; Tannenbaum et al. 1978; Fishbein 1979; Sander 1982), formaldehyde and chloral (Kefer and Roller 1973), phenols (Walker et al. 1979) and fulvic acid (Weerasooriya and Dissanayake 1989) as well as by photochemical reactions (Ohta et al. 1982).

The acceleration of nitrosomorpholine formation by thiocyanate was studied in detail by Fan and Tannenbaum (1973). Thiocyanates and halides produce the nitrosating species ON-X (X = NCS, Br, Cl) which is more reactive than $ON\text{-}NO_2$. This thiocyanate- or halide-catalyzed nitrosation of amines is especially important at low pH values (pH < 2.5), at high thiocyanate or halide concentrations, and at low nitrite concentrations (Fan and Tannenbaum 1973; Mirvish 1975). Fan and Tannenbaum (1973) give an order of acceleration effectiveness for nitrosomorpholine formation being $SCN^- >> Br^- >> Cl^- > SO_4^{2-} = ClO_4^- = H_2PO_4^- = RCO_2^-$. They also give rate constants for the nitrosation of morpholine in the presence of thiocyanate and bromide. They found an acceleration factor for nitrosomorpholine formation of approximately 10 for bromide and of approximately 500 for thiocyanate. Fan and Tannenbaum (1973) also found that the optimum pH for nitrosation is shifted by the anions under investigation. Wainright (1986) reports an order of catalytic reactivity for nitrosamine formation being $I^- > SCN^- > Br^- > Cl^-$ with thiocyanate accelerating the reaction 20fold. The mechanism of the catalytic acceleration of nitrosation can be explained by the equations given in Figure 3.2. By reaction of nitrous acid with thiocyanate (or halides) the respective nitroso thiocyanates (nitroso halides) are formed. These are more effective nitrosating agents as the nitrosyl cation or N_2O_3 and thus enhance the subsequent nitrosamine formation.

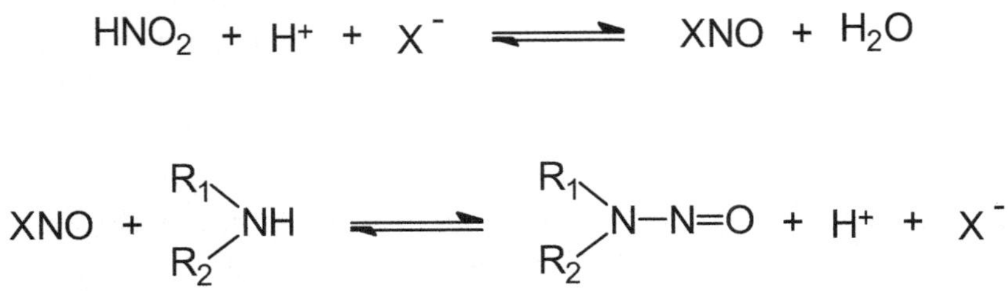

X = SCN, Cl, Br

Source: Fan and Tannenbaum (1973)

Figure 3.2 Catalytic acceleration of nitrosamine formation by thiocyanate and halides

Reducing agents like ascorbic acid, on the other hand, can minimize or even prevent nitrosamine formation by reduction of nitrite to nitric oxide (Mirvish et al. 1972; Mirvish 1975; Fishbein 1979; Preussmann 1984; Williams 1994; Kolb et al. 1997). Gallic acid, tannin, cysteine, and sulfite also proved to be able to minimize nitrosamine formation but were found to be less effective than ascorbic acid at all pH values (Fishbein 1979). Mirvish found that urea and ammonium sulfamate are less effective agents for blocking the nitrosamine formation of morpholine, piperazine, N-methylaniline, methylurea, and dimethylamine (Mirvish et al. 1972). Loeppky et al. (1994) studied in detail various possibilities for blocking nitrosamine formation by nitrosation. They found that for instance pyrrole is very effective in blocking the nitrosation of secondary amines because it reacts very quickly with the nitrosating agent (without forming nitrosamines). Monohydroxy phenols catalyze nitrosamine formation (Walker et al. 1979), but polyhydroxy phenols like hydroquinone or catechol are very effective in blocking nitrosation (Loeppky et al. 1994). The results of Loeppky et al. (1994) for the ability of various compounds to block N-nitrosomorpholine formation are summarized in Table 3.4.

Loeppky et al. (1994) even tried to immobilize pyrroles and hydroquinones by incorporating them into polymers and other solid matrices. They could demonstrate that all of the materials prepared were effective blocking agents.

Table 3.4
Ability of various compounds to block N-nitrosomorpholine formation

97 – 88% Block	87 – 80% Block	75 – 63% Block
1-Benzyl-2,5-dimethylpyrrole	4-Methylcatechol	Hydrazine
2,5-Dimethylpyrrole	Hydroquinone	3,4-Diaminobenzoic acid
Pyrrogalol	Catechol	Octamethylporphine
Phloroglucinol	1-Benzylpyrrole	1,2,5-Tribenzylpyrrole
Pyrrole	Gallic acid	Ammonium sulfamate
	Ascorbic acid	2,5-Diphenylpyrrole
	p-Phenylenediamine	2-Phenyl-3-benzoylaziridine
	o-Phenylenediamine	
	4-Ethylaniline	

Source: Loeppky et al. (1994)

In his review, Wainright (1986) also gives a list of compounds inhibiting nitrosamine formation by competing for the nitrosating agents, including phenols, alpha-tocopherol and other aromatic or heterocyclic ring compounds, primary amines, amides and peptides, purines, thio compounds including proteins, lipids and unsaturated compounds, ascorbic acid, and small hydroxyl compounds including sugars and alcohols.

The role of microorganisms in nitrosamine formation in the environment is not completely elucidated. Some authors (Ayanaba and Alexander 1974; Archer et al. 1978; Kobayashi and Tchan 1978; Greene et al. 1981; Yordy and Alexander 1981) assume that microorganisms play a significant role in nitrosamine formation in natural waters and in soil, whereas others conclude from their studies that microorganisms are not directly involved in nitrosamine formation (Mills and Alexander 1976). Microorganisms, however, might enhance the availability of possible precursor compounds. Hawksworth and Hill (1971) found that five out of ten strains of *Escherichia coli* tested were able to catalyze nitrosamine formation from diphenylamine, dimethylamine, piperidine, pyrrolidine and N-methylaniline even at pH 6.5, where in absence of the bacteria no nitrosamine formation was observed. They attributed this accelerating effect mainly to a nitrate reductase activity of the bacteria. A catalytic acceleration of nitrosamine formation by bacteria was also found by Leach et al. (1987).

Most chemical reactions, including nitrosamine formation, tend to proceed faster when the temperature is raised. Nevertheless, nitrosation is usually very rapid even at low temperatures and, thus, temperature control seems to be no proper option for minimizing nitrosamine formation (Wainright 1986). Fan and Tannenbaum (1973) calculated the activation energy for the nitrosation of morpholine to approximately 10 kcal/mol.

Wilczak et al. (2004) studied the effect of nitrification on NDMA formation in distribution systems after chloramination and found no evidence that nitrite formed in the distribution system due to nitrification significantly contributes to additional NDMA formation in the treated water.

Choi and Valentine (2003) introduced another possible pathway for NDMA formation by nitrosation of dimethylamine involving a reaction of dimethylamine with nitrite that is catalyzed or enhanced by free chlorine. They predict that through oxidation of nitrite by free chlorine dinitrogen tetroxide (N_2O_4) is formed as an intermediate, being a very effective nitrosating agent. The formation of nitrogen tetroxide from nitrite is favorable at neutral pH (in contrast to the formation of N_2O_3 which is enhanced at low pH). Figure 3.3 summarizes this reaction pathway. Only the dinitrogen tetroxide species $ON\text{-}NO_3$ is able to form nitrosamines, whereas the species $O_2N\text{-}NO_2$ favors the formation of the respective nitrates (Wainright 1986).

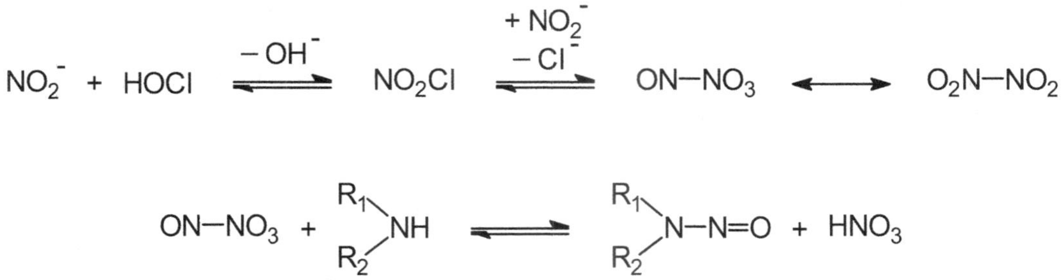

R_1, R_2: alkyl groups, e.g. methyl or ethyl

Source: Choi and Valentine (2003)

Figure 3.3 Formation of nitrosamines by chlorine enhanced nitrosation of secondary amines

Nitrosamine Formation by Reaction of Inorganic Chloramines with Aliphatic Amines

The UDMH Pathway

In addition to nitrosamine formation by nitrosation of amines, another reaction pathway for NDMA formation was proposed which seems to be especially relevant for nitrosamine formation during disinfection of waste and drinking waters with chloramine (Choi and Valentine 2002a, 2002b; Choi et al. 2002; Mitch and Sedlak 2002b; Mitch et al. 2003a). This mechanism involves the reaction of monochloramine with dimethylamine to unsymmetrical dimethylhydrazine (UDMH; 1,1-dimethylhydrazine) followed by oxidation of UDMH to NDMA. The proposed mechanism accounts for the fact that NDMA formation has been observed during disinfection with chlorine and chloramines even in the absence of nitrite.

The use of chloramine for disinfection of waste and drinking waters has a long time tradition in the United States as well as in Europe (Haase 1936). Due to taste and odor problems related to the use of chloramine, however, the German Drinking Water Directive does not allow the use of chloramine for disinfection. Chloramines are present in aqueous solutions if free chlorine and ammonia are applied for disinfection in molar dose ratios of less than one. At pH levels usually present in drinking waters, the predominant chloramine species are monochloramine (NH_2Cl) and dichloramine ($NHCl_2$) (Schreiber and Mitch 2005). Diyamandoglu and coworkers (Diyamandoglu and Selleck 1992; Diyamandoglu 1994) could prove that nitrite formation during chloramination can be ruled out and thus another mechanism of nitrosamine formation must account for the observed occurrence of NDMA in drinking waters disinfected by chlorine and chloramine.

Choi and Valentine (2002a) found in laboratory-based experiments that even if nitrite is present during chloramination, the NDMA levels in the disinfected water are much higher than could be expected from pure nitrosation of dimethylamine. In their experiments they also found that NDMA was formed when only chlorine was added to an aqueous solution of dimethylamine. After addition of ammonia, however, the amount of NDMA formed was much higher. Experiments from Najm and Trussell (2001), on the other hand, indicate that nitrosamine formation during disinfection with chlorine requires the presence of ammonia (leading to the formation of chloramines). The sole application of ozone did not cause any nitrosamine formation (Najm and Trussell 2001).

The reaction of monochloramine and dimethylamine (DMA) was used in former times in the so-called Raschig process for manufacturing of UDMH, a component of liquid rocket fuels (see Figure 3.4) (Choi and Valentine 2002a, 2002b; Mitch et al. 2003a). In most cases, UDMH formation from amines is quite slow and can even take days. Choi and Valentine (2002a), however, observed a catalytic effect of bromide on NDMA formation which can be attributed to the higher reactivity of bromamines compared to chloramines.

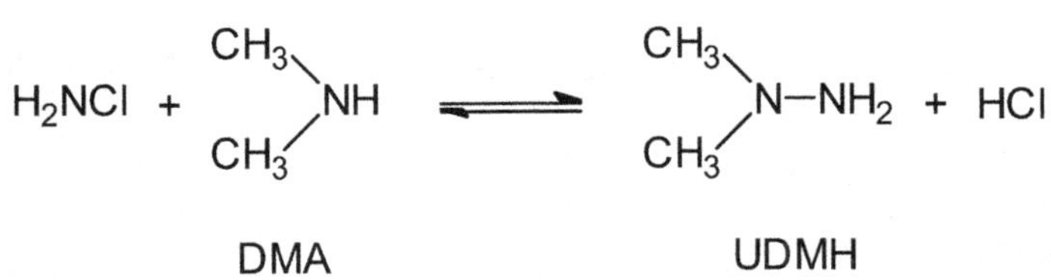

Figure 3.4 Formation of unsymmetrical dimethylhydrazine (UDMH) from monochloramine and dimethylamine (DMA)

The formation of NDMA by oxidation of UDMH was already observed in the mid 1980s during treatment of UDMH-containing rocket fuels by hypochlorite (Mitch et al. 2003a). Other oxidants which proved to be capable of transforming UDMH to NDMA include cupric ion (Banerjee et al. 1984), potassium permanganate, iodate (Castegnaro et al. 1986), hydrogen peroxide, and oxygen (Lunn et al. 1991; Lunn and Sansone 1994). When UDMH used in rocket fuels was deliberately exposed to the atmosphere, the formation of NDMA was observed and explained by the oxidation of UDMH by oxygen (Lunn et al. 1991; Lunn and Sansone 1994). Optimum conditions for NDMA formation from UDMH were identified to be a neutral to slightly basic pH (Lunn et al. 1991; Mitch and Sedlak 2002b). In addition to NDMA, other products are formed by oxidation of UDMH. Mitch and Sedlak (2002b) mention the formation of dimethylcyanamide, dimethylformamide, and formaldehyde monomethylhydrazone. In laboratory-based experiments they could verify the formation of dimethylcyanamide and dimethylformamide during chloramination of dimethylamine solutions.

The formation of NDMA by reaction of monochloramine with aliphatic amines during chlorination or chloramination of waste and drinking water was studied in more detail by Valentine, Mitch and their coworkers (Choi and Valentine 2002a, 2002b; Choi et al. 2002; Mitch and Sedlak 2002b). The key reactions of the multi-step mechanism proposed by Choi and Valentine (2002a, 2002b) are given in Figure 3.5 for the reaction between DMA and monochloramine, whereby side-reactions and by-products have been omitted in order to simplify the presentation.

In chlorinated waters containing ammonia, first, monochloramine is formed by reaction of hypochlorous acid (or any hypochlorite salt) and ammonia (1). Monochloramine then reacts with DMA by formation of UDMH (4). DMA may also combine with either hypochlorous acid (2) or monochloramine (3) to form dimethylchloramine (DMCA) which cannot react to UDMH. These side-reactions thus limit the amount of DMA available for NDMA formation. Rate constants of the side-reactions, however, are low compared to the rate constants of monochloramine formation. In a final step, UDMH is oxidized by monochloramine to NDMA (5). Choi and Valentine (2002a) assume that this oxidation is preferentially done by monochloramine. Oxidation of UDMH by hypochlorite, however, cannot be excluded (Castegnaro et al. 1986).

$$HOCl + NH_3 \rightleftharpoons NH_2Cl + H_2O \qquad (1)$$

$$HOCl + \underset{\text{DMA}}{(CH_3)_2NH} \rightleftharpoons \underset{\text{DMCA}}{(CH_3)_2NCl} + H_2O \qquad (2)$$

$$NH_2Cl + \underset{\text{DMA}}{(CH_3)_2NH} \rightleftharpoons \underset{\text{DMCA}}{(CH_3)_2NCl} + NH_3 \qquad (3)$$

$$\underset{\text{DMA}}{(CH_3)_2NH} + NH_2Cl \underset{(4)}{\overset{OH^-}{\rightleftharpoons}} \underset{\text{UDMH}}{(CH_3)_2N{-}NH_2} \underset{(5)}{\overset{\text{oxidation}}{\rightleftharpoons}} \underset{\text{NDMA}}{(CH_3)_2N{-}N{=}O}$$

Source: Choi and Valentine (2002a, 2002b)

Figure 3.5 Formation of NDMA in chlorinated waters

Valentine and coworkers (Choi and Valentine 2002a, 2002b; Choi et al. 2002) also developed a kinetic model to validate their proposed mechanism. Assuming that all reactions except for the final oxidation of UDMH by monochloramine are elementary and can be described by second order reactions, and by using either reaction constants from literature or data from their own experiments, they could quite adequately predict the formation of NDMA from DMA and monochloramine in several laboratory-scale experiments. For the oxidation of UDMH they assume that the rate limiting reaction is first order in both DMA and monochloramine concentration.

The model of Choi and Valentine predicts a plateau in NDMA formation with increasing monochloramine concentration due to the competing formation of DMCA. A maximum in NDMA formation is expected for a ratio of DMA to monochloramine concentration of approximately one. These predictions are in good accordance with the results of their laboratory-scale experiments (Choi and Valentine 2002a, 2002b). Another conclusion from this model is that in the presence of excess monochloramine, NDMA formation should be approximately proportionnal to the concentration of organic nitrogen-containing precursors (Choi and Valentine 2002a). Najm and Trussell (2001), however, found in their studies that NDMA formation strongly depends on the dose of disinfectant but not very much on the NOM content or the nitrogen content of the water.

Based on the fact that hypochlorite formed an order of magnitude more NDMA than monochloramine when applied to secondary municipal wastewater effluents that contained excess ammonia and where thus monochloramine was expected to be the predominant oxidant, Mitch and coworkers reinvestigated the mechanisms that might have an impact on NDMA formation during chloramination (Schreiber and Mitch 2005; Mitch et al. 2005). Depending on the ratio of chlorine to ammonia (Cl:N ratio) and on pH, three different inorganic chloramines can be formed at different proportions: monochloramine (NH_2Cl), dichloramine ($NHCl_2$) and trichloramine (NCl_3), with a maximum of two inorganic chloramines coexisting at specific conditions. While the monochloramine fraction dominates at molar Cl:N ratios < 1, the formation of dichloramine increases with increasing Cl:N ratio. At Cl:N ratios > 1.5 trichloramine, dichloramine and free chlorine coexist. Another strong determinant of chloramine speciation is pH. At a molar Cl:N ratio of 0.85, the average percentage contribution of dichloramine ($NHCl_2$) to the total oxidant concentration was reported to decline from 50% at pH 5.1, to 15% at pH 6.9, to negligible levels at pH 8.8 (Schreiber and Mitch 2005). It was further found that the rate of NDMA formation from the reaction of dichloramine with organic nitrogen-containing precursors is much faster than the analogous reactions involving monochloramine. In laboratory-based experiments, for example, an order of magnitude more NDMA was formed from $NHCl_2$/dimethylamine than from NH_2Cl/dimethylamine (Schreiber and Mitch 2005). These results indicate, that NDMA formation during chloramination is enhanced in the presence of dichloramine ($NHCl_2$), with molar Cl:N ratios and pH values being important determinants for the amount of NDMA formed. Although hypochlorite usually is added in average Cl:N ratios < 1 during chlorination of ammonia-containing wastewater or chloramination of drinking water, the molar Cl:N rate at the point of reagent addition prior to complete mixing is very different from that of the entire solution after mixing. Thus, in cases where hypochlorite is applied to a well-mixed solution of NH_4Cl, very high Cl:N ratios might exist at the point of oxidant addition that cause locally formation of dichloramine. Once formed, dichloramine is fairly stable, even upon reaction with residual ammonia after mixing (Schreiber and Mitch 2005; Mitch et al. 2005). The research results presented above imply that one strategy to minimize NDMA formation during chloramination is the production of chloramines in a manner that minimizes the concentration of dichloramine, i.e. addition of preformed chloramine (prepared separately in a dosing tank at elevated pH) or addition of free chlorine prior to ammonia.

Furthermore, these and other investigations revealed another factor being important for nitrosamine formation, i.e. the chlorination state of organic nitrogen precursors (Mitch and Sedlak 2002b; Wilczak et al. 2003; Schreiber and Mitch 2005; Mitch et al. 2005). The chlorination of organic nitrogen species lowers the formation of NDMA because the reaction of chlorinated NDMA precursors with chloramines does not result in any significant production of NDMA (Mitch and Sedlak 2002b). Laboratory-based experiments at neutral pH values showed that NDMA formation from the interaction of preformed monochloramine and dichloramine with chlorinated dimethylamine was nearly an order of magnitude less than from the corresponding reactions with dimethylamine (Schreiber and Mitch 2005). In agreement with these findings, Wilczak et al. (2003) reported that a change from chloramination to free chlorine disinfection followed by chloramination significantly reduced NDMA formation at a drinking water treatment plant where cationic polymers served as the source of organic nitrogen-containing NDMA precursors. Thus, in order to decrease NDMA formation, chlorine addition prior to addition of ammonia could be employed to chlorinate the organic nitrogen-containing species in fully nitrified wastewaters or in drinking waters (Schreiber and Mitch 2005). However, in this case, the formation of free chlorine disinfection by-products has to be minimized by an optimal chlorine dosage (Mitch et al. 2005). In the case of non-nitrified wastewater effluents, application of preformed monochloramine may be advantageous. However, even though direct application of hypochlorite to wastewaters containing excess ammonia increases NDMA formation, the magnitude of this increase might be lower than expected due to partial chlorination of organic nitrogen precursors. The rate constant for chlorine transfer from hypochlorous acid to dimethylamine ($6.1 \times 10^7\ M^{-1}s^{-1}$) is higher than the corresponding reaction with ammonia ($4.2 \times 10^6\ M^{-1}s^{-1}$) (Mitch et al. 2005).

In the experiments reported above, the investigators focused mainly on DMA as precursor for NDMA formation. However, in analogy to the findings for nitrosamine formation via the nitrosation mechanism, several authors also documented that NDMA during chloramination is not only formed from dimethylamine, but also from diverse tertiary amines with dimethylamine functional group (Mitch and Sedlak 2002a, 2004). Studies comparing the NDMA formation potential with the dimethylamine concentrations in different surface and waste waters revealed that only a small portion (< 25%) of the formed NDMA could be attributed to dimethylamine (Gerecke and Sedlak 2003; Mitch et al. 2003b).

In bench-scale tests, Mitch and Sedlak (2004) demonstrated the formation of NDMA from tertiary amines with a dimethylamine functional group (e.g. trimethylamine) that was only slightly lower than from dimethylamine. Amines without dimethylamine functional group did not form significant amounts of NDMA. They also proposed a mechanism for NDMA formation from tertiary amines with a dimethylamine functional group. The initial step of this mechanism is a dealkylation of the tertiary amine during chloramination resulting in the formation of dimethylamine which then is transformed to NDMA. According to the authors, an iminium ion is formed as intermediate of the dealkylation step.

Mitch and Sedlak (2002b) also investigated the formation of other nitrosamines apart from NDMA during chloramination. After chloramination of the effluents of three wastewater treatment plants they were able to detect NDMA as well as nitrosodiethylamine (NDEA), N-nitrosopyrrolidine (NPYR), and N-nitrosopiperidine (NPIP). In all of their experiments, however, NDMA concentrations were much higher than the concentration levels of the other nitrosamines. NDEA and NPYR were formed at one order of magnitude lower concentrations than NDMA, while NPIP was formed at two orders of magnitude lower concentrations.

The Chlorinated UDMH Pathway

In a recent paper, Schreiber and Mitch describe another pathway that leads to the formation of nitrosamines during chloramination (Schreiber and Mitch 2006b). For the formation of NDMA, the key reactions of this alternative pathway are given in Figure 3.6. Based on results of laboratory-scale experiments, Schreiber and Mitch propose that the initial step involves a nucleophilic substitution reaction between secondary amines (e.g., dimethylamine) and dichloramine to form chlorinated unsymmetrical dialkylhydrazine intermediates (e.g, chlorinated UDMH for DMA), rather than unsymmetrical dialkylhydrazine intermediates (e.g., UDMH). These intermediates can then be oxidized to the respective nitrosamine or other yet unidentified products, either by dissolved oxygen or by chloramines. Although monochloramine is the predominant chloramine species under typical chloramination conditions, the model of Schreiber and Mitch predicts NDMA concentrations formed during laboratory-scale experiments based solely on the low levels of dichloramine formed via disproportionation of monochloramine (Schreiber and Mitch 2006).

The justification of Schreiber and Mitch for chlorinated UDMH as key-intermediate for NDMA formation during chloramination is based on two experimental facts: First, the rate constant for the formation of the chlorinated UDMH according to Figure 3.6 is three orders of magnitude faster than the rate constant for the formation of UDMH from monochloramine and dimethylamine. Second, the structure of chlorinated UDMH enables the incorporation of dissolved oxygen and thus chlorinated UDMH can be oxidized to NDMA by dissolved oxygen. The oxidation of UDMH by dissolved oxygen on the other hand requires the presence of metals catalyzing the electron transfer process. As a consequence, NDMA formation during chloramination could be reduced by reducing the concentration of dissolved oxygen (Schreiber and Mitch 2006b).

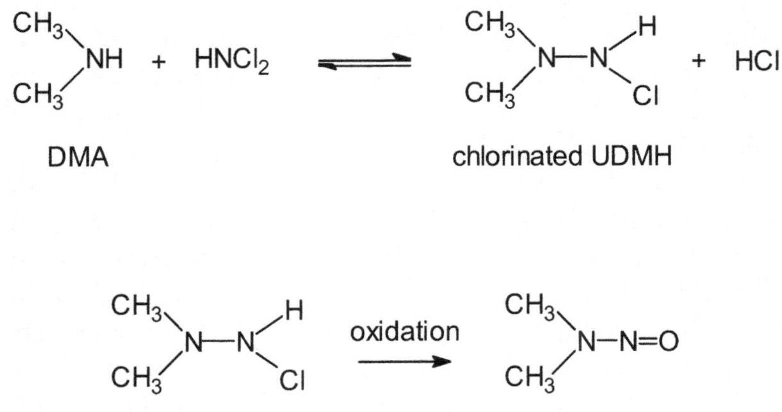

Source: Schreiber and Mitch (2006b)

Figure 3.6 Formation of NDMA via chlorinated UDMH

Up to now, however, neither UDMH nor chlorinated UDMH could be unambiguously identified, by analytical techniques such as LC-MS.

VALIDATION OF THE NFP TEST

In order to properly validate the NFP test as described in Chapter 2, the impact of several experimental parameters such as reaction time, pH, and chloramine dose on the levels of nitrosamines formed was determined. Furthermore, repeatability and reproducibility of the test procedure were determined. Those validation experiments were performed with natural waters of different origin and with different characteristics. The main physical-chemical parameters characterizing those waters are summarized in Table 3.5. The tap water was produced from a deep groundwater (no anthropogenic impact) and treated by aeration and sand filtration (no disinfection).

Impact of Reaction Time

To investigate the formation of nitrosamines with time, a batch of NFP tests was performed with river water A (river water influenced by industrial and domestic waste waters at a medium level; see also Table 3.5). Each test was performed in a separate bottle and in each bottle the chloramine was quenched by addition of sodium sulfite after a defined period of time. After nitrosamine analysis, the amounts of NDMA, NDEA, and NPYR after defined reaction times, i.e. defined contact times between water sample and chloramine, were obtained. The results of these experiments are presented in Figure 3.7.

Table 3.5
Physical-chemical characteristics of natural waters under investigation

Water sample (sampling date)	pH	Alkalinity (mg/L $CaCO_3$)	Ammonia (mg/L NH_4^+)	Nitrate (mg/L NO_3^-)	Nitrite (mg/L NO_2^-)	TOC (mg/L)	UV_{254} (m^{-1})
River water A (10.09.2004)	8.04	117	0.01	5.1	0.04	1.6	3.8
River water B (09.09.2004)	8.14	125	0.01	8.2	< 0.01	1.9	5.2
River water C (27.11.2003)	7.50	25	< 0.01	7.2	< 0.01	2.3	6.2
Groundwater A (28.09.2004)	n.d.	266	< 0.01	8.3	< 0.01	1.3	2.1
Groundwater B (27.09.2004)	n.d.	437	< 0.01	< 0.5	< 0.01	4.5	12.2
Tap water A (22.09.2004)	7.45	260	< 0.01	5.8	< 0.01	0.88	1.1

n.d.: not determined

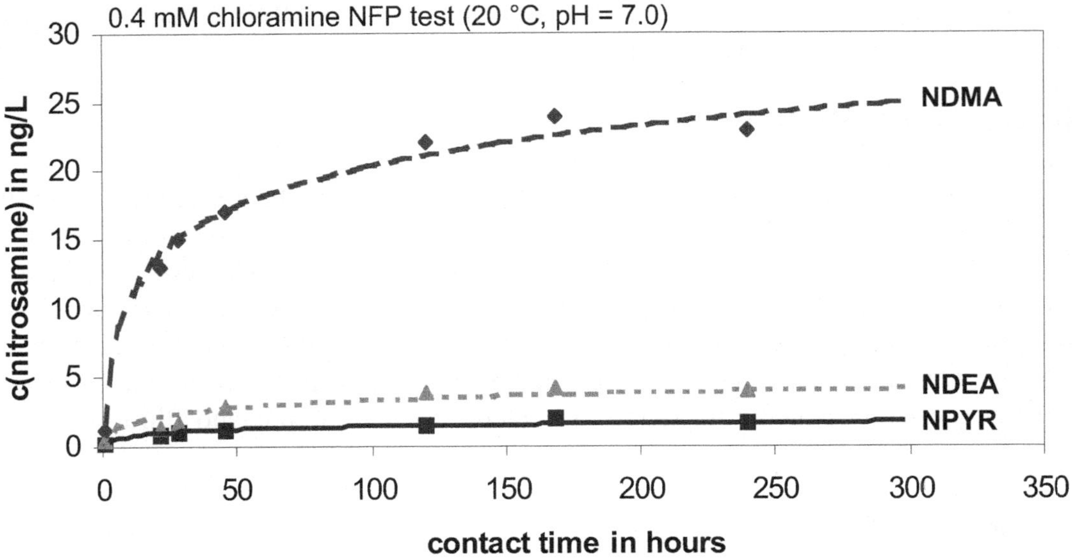

Figure 3.7 Nitrosamine formation in the NFP test with river water A as a function of time

Obviously, levels of NDEA and NPYR were about one order of magnitude lower than NDMA levels. However, the time-dependent courses of nitrosamine formation were comparable for all of the nitrosamines, all reaching a plateau after 150-200 hours of chloramine contact time. Although it is not possible to generalize the result of one experiment and time-dependence of nitrosamine formation may depend on several factors, as age and origin of natural organic matter or degree and nature of anthropogenic impact, the 168 hour (7 day) contact time chosen for the NFP test seems to be reasonable to achieve almost maximum nitrosamine levels within an acceptable period of time.

Impact of pH

For investigating the impact of pH on nitrosamine formation in the NFP test, the natural waters characterized in Table 3.5 were adjusted to different pH values by addition of phosphate buffer solutions with varying hydrogenphosphate:dihydrogenphosphate concentration ratios. As in most of the waters, only NDMA was formed in relevant amounts, only these results were evaluated. A concise presentation of the data is given in Table 3.6. Figure 3.8 gives additional data for 10 river water samples which have been tested in the NFP test at pH values of 7 and 8.

Table 3.6
NDMA formation in natural waters in NFP tests at different pH values

	pH		
Water sample	7.0	7.5	8.0
River water 1	25	35	36
River water 2	-	29	26
Groundwater 1	5.9	4.5	-
Groundwater 2	-	8.2	7.6
Tap water 3	-	1.8	2.9

NDMA concentrations in ng/L; 20°C, 0.4 mM chloramine

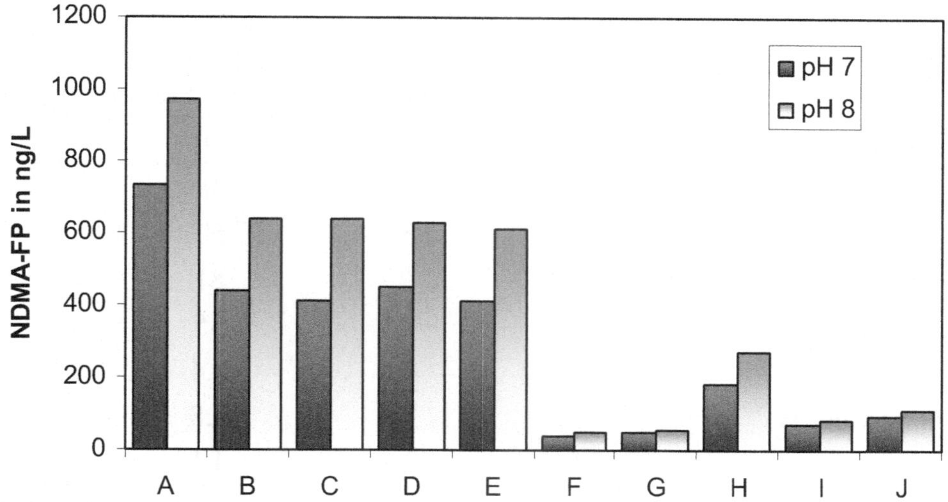

Conditions: 20°C, 0.4 mM chloramine

Figure 3.8 NDMA formation in river waters in NFP tests at different pH values

For some of the waters, there seems to be a significant increase of NDMA concentration with increasing pH, whereas for others NDMA formation in the NFP test seems not to depend on pH. The data in Table 3.6 and Figure 3.8 indicate that especially for samples showing a high NDMA formation potential, the effect of pH becomes significant, leading to a higher nitrosamine formation at higher pH values. From the data it can be expected that for some waters at a pH of 9 even higher nitrosamine levels could be found. However, no clear correlation of pH effects to general water characteristics (as age of groundwater or degree of anthropogenic pollution) could be found. Because for some waters (especially those exhibiting a high alkalinity) precipitations occurred when pH was adjusted to 8, it was decided to perform all future NFP tests at a pH of 7, even if a slightly higher NFP could be expected at a higher pH.

Impact of Chloramine Dose

To investigate the impact of the amount of chloramine used in the NFP test on the levels of nitrosamines, the natural waters used in previous experiments (see Table 3.5) were treated with different amounts of monochloramine solution. To check, whether there is an interference with pH effects, different pH values were applied at some of the experiments.

Again, for most of the waters, only NDMA was formed in relevant amounts. The results are presented in Table 3.7. They prove that there is an impact of chloramine dose on NDMA formation in the NFP test. For most of the natural waters, an increase in NDMA formation is observed with increasing chloramine dose. The effect seems to be more pronounced if only small amounts of NDMA are formed (e.g. for groundwater A and tap water A). If higher amounts of NDMA are formed (e.g. for the river waters, especially at higher pH values), the impact of the chloramine dose seems to become negligible. Thus, for most of the further tests, chloramine doses of 0.4 mM (= 30 mg/L as Cl_2) were applied.

Table 3.7
NDMA formation in natural waters in NFP tests with different chloramine doses

		Total chlorine concentration (mg/L Cl_2)		
Water sample	pH	30	70	140
River water A	7.0	33	38	48
River water A	8.0	55	57	56
River water B	8.0	79	81	88
River water C	8.3	26	35	-
Groundwater A	7.5	4.6	8.4	-
Groundwater B	7.2	7.6	9.8	-
Tap water A	7.6	2.9	4.1	-

NDMA concentrations in ng/L; 20°C

Repeatability and Reproducibility of the NFP Test

To complete the validation of the NFP test, repeatability (in-day variance) and reproducibility (between-day variance) of the test procedure were determined. For these purposes, a NFP test with two river waters (A and B) was performed in triplicate. After two weeks, the procedure was repeated with a freshly prepared chloramine stock solution but with the same river waters which were stored in between at 4 °C in the dark. The results of these NFP tests for NDMA, NDEA, and NPYR are given in Figures 3.9 to 3.13. Other nitrosamines were not formed in detectable amounts.

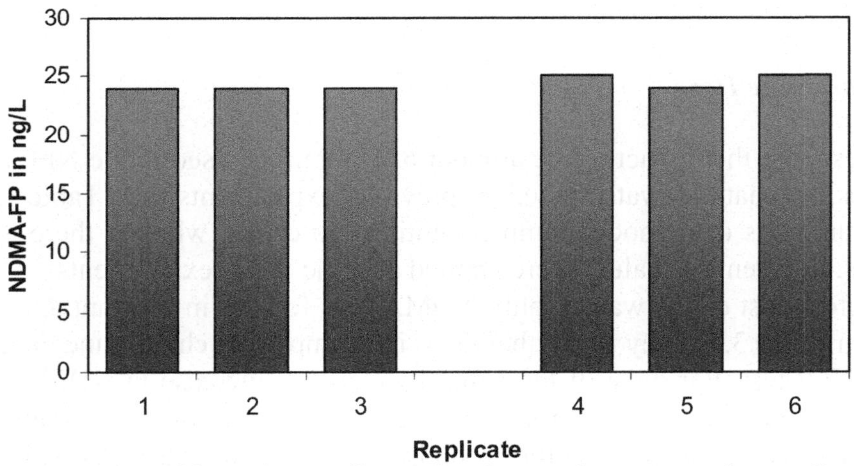

Conditions: pH = 7.0, 0.4 mM chloramine; replicates 1 to 3 were done on one day, replicates 4 to 6 two weeks later

Figure 3.9 NDMA formation in NFP tests with river water A

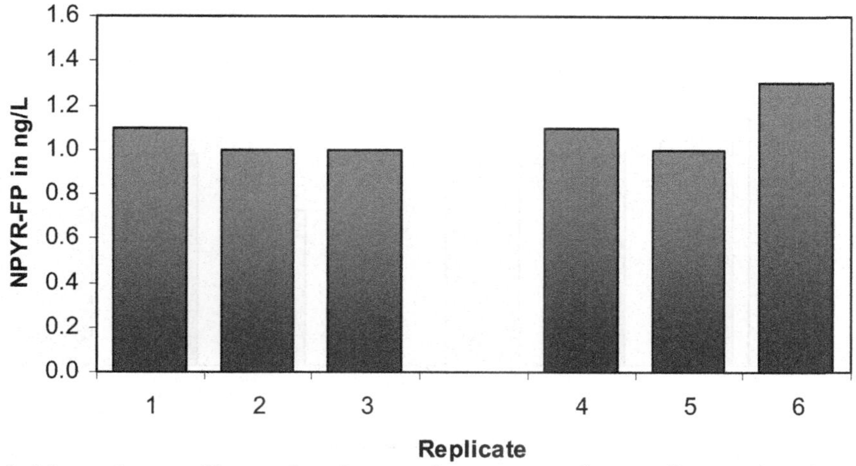

pH = 7.0, 0.4 mM chloramine; replicates 1 to 3 were done on one day, replicates 4 to 6 two weeks later

Figure 3.10 NPYR formation in NFP tests with river water A

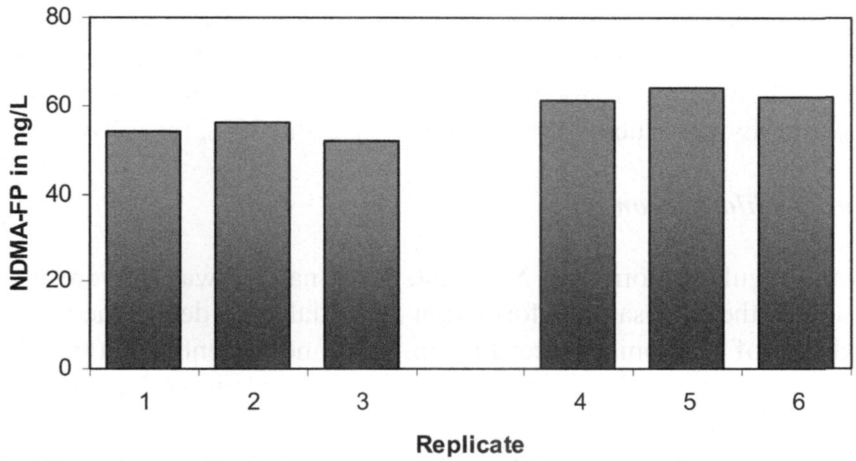

pH = 7.0, 0.4 mM chloramine; replicates 1 to 3 were done on one day, replicates 4 to 6 two weeks later

Figure 3.11 NDMA formation in NFP tests with river water B

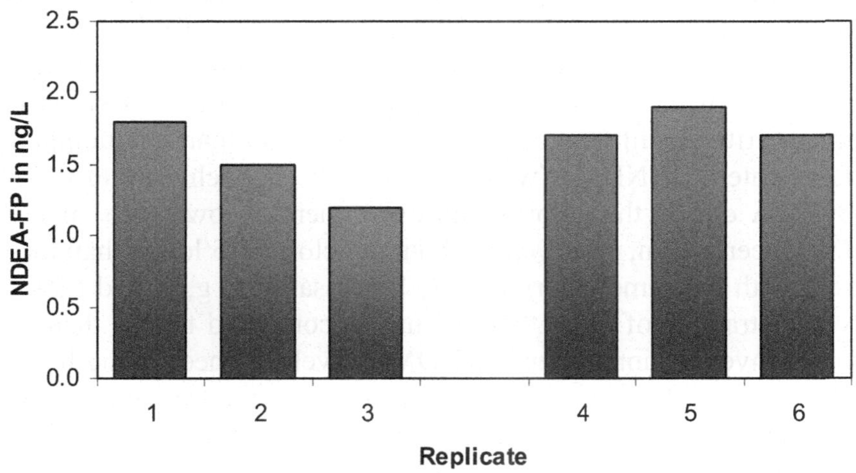

pH = 7.0, 0.4 mM chloramine; replicates 1 to 3 were done on one day, replicates 4 to 6 two weeks later

Figure 3.12 NDEA formation in NFP tests with river water B

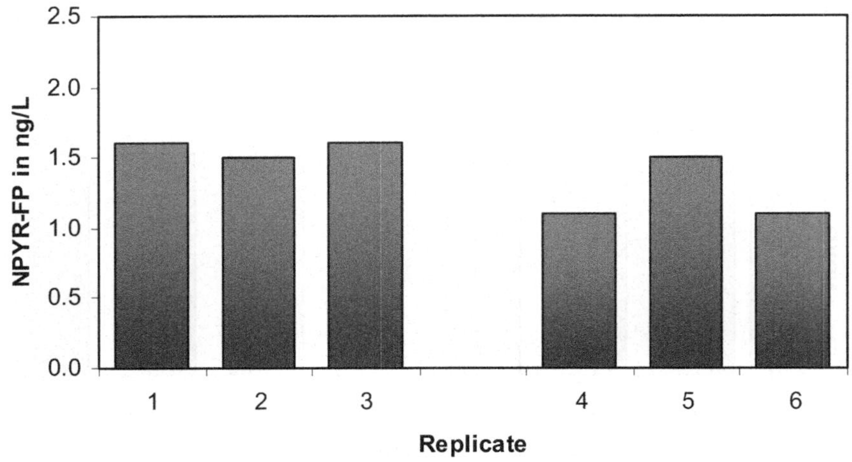

pH = 7.0, 0.4 mM chloramine; replicates 1 to 3 were done on one day, replicates 4 to 6 two weeks later

Figure 3.13 NPYR formation in NFP tests with river water B

The data show that the repeatability of the overall process including NFP test and analysis of nitrosamines is good. Depending on the concentration levels of the nitrosamines, within-day repeatability is between 5% (for high nitrosamine levels) and 20% (for low nitrosamine levels) and between-day reproducibility is between 10% and 20%, respectively.

Chloramination vs. Chlorination

Besides the regular chloramine NFP test, some natural waters were investigated with a chlorination test, i.e. the nitrosamine formation potential was determined during chlorination without any addition of ammonia. According to Choi and Valentine (2003), NDMA might be formed during chlorination with hypochlorite through the oxidation of nitrite via dinitrogen tetroxide (N_2O_4) as an intermediate.

In total, nine river water samples with different levels of anthropogenic contamination were tested both with the chloramination test and with the chlorination test. One of these samples was further investigated with both tests at three different pH values (7.0; 7.5; 8.0) and at two dosages of chlorination agent (30 mg/L and 70 mg/L Cl_2 at pH 7.0). Figures 3.14 and 3.15 give the results of these experiments for NDMA (other nitrosamines were not found).

From the results of these experiments, it becomes obvious that chlorination with hypochlorite is much less critical with regard to NDMA formation than chloramination. In six of the nine tested surface waters no NDMA was formed in the hypochlorite test. Only three surface waters formed NDMA during the chlorination experiment at low levels of less than 2.5 ng/L. Even this NDMA concentration, however, is about a factor of 35 lower than the one formed during chloramination with the same chlorine dose. Since samples g, h, and i also showed relative high ammonia concentrations of more than 0.1 mg/L (compared to less than 0.05 mg/L for the other samples under investigation), even the NDMA levels formed during hypochlorite application might be a result of a chloramination mechanism. Thus, the formation of NDMA via the dinitrogen tetroxide pathway during chlorination seems to be much less significant compared to NDMA formation during chloramination.

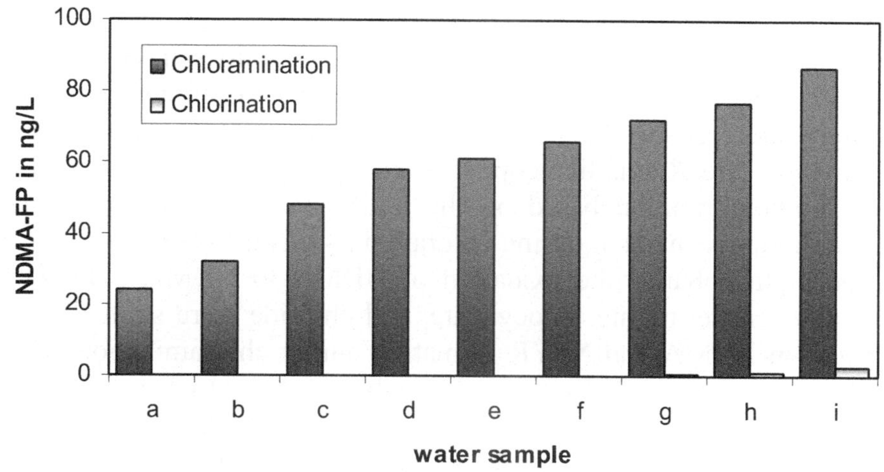

pH = 7.0, 30 mg/L chlorine, 168 h contact time

Figure 3.14 Comparison of the NDMA formation potential during chloramination and chlorination of river water samples

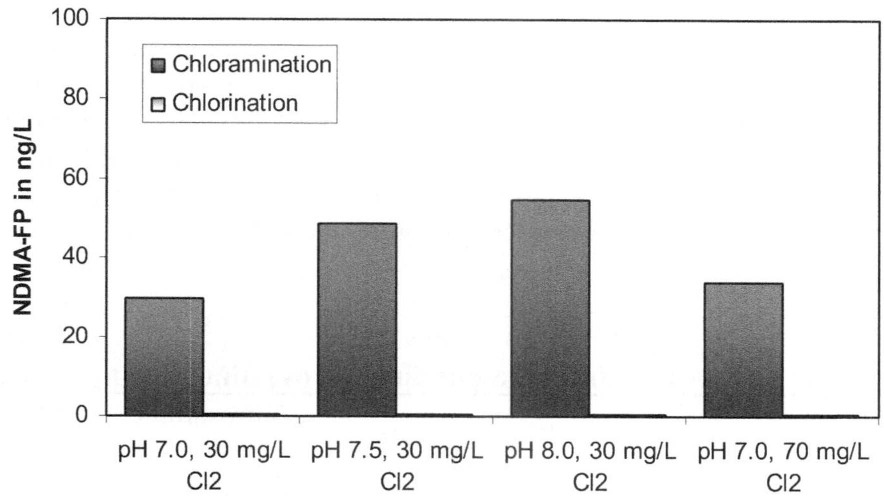

Figure 3.15 Comparison of the NDMA formation potential during chloramination and chlorination of one river water sample applying different experimental conditions

Chloramination: Conversion Rates of Secondary Amines and Catalytic Effects

In order to determine conversion rates of the parent amines into the corresponding nitrosamines during chloramination, drinking and surface waters were spiked with secondary amines at a concentration of either 1,000 ng/L or 10,000 ng/L and were then chloraminated using standard conditions. The results of those experiments are presented in Table 3.8. It can be seen that the percentage of conversion based on the molar concentrations is generally rather low, ranging from 0.02 to 2.74% and depending on the amine tested and its concentration. Conversion rates of DMA tend to be slightly higher in surface than in drinking water and at higher amine concentrations. The results for DMA/NDMA conversion (0.49-2.74%) are well comparable to former findings reported in literature. Starting with a solution of 73 µg/L of DMA in deionized water, the application of the NFP test by Mitch et al. (2003) revealed that about 0.5% of DMA were converted to NDMA. Similar results were obtained by Gerecke and Sedlak (2003) who spiked surface water with DMA and obtained a yield of about 0.6% NDMA.

Comparing conversion rates of secondary precursor amines and their concentration levels found in environmental samples as well as the amount of nitrosamines formed during disinfection in U.S. waterworks leads to the conclusion that there might be additional effects enhancing nitrosamine formation. Metal ions like copper, iron, and nickel were shown in earlier studies to accelerate nitrosamine formation based on the traditional nitrosation pathway (Challis et al. 1978). Even though the mechanism during chloramination is different from this nitrosation pathway, metal ions might enhance the oxidation of UDMH to NDMA (Choudhary and Hansen 1998). Additionally, nitrite, nitrate, thiocyanate, and chloride were studied with regard to their catalytic influence on NDMA and NPYR formation during chloramination. Therefore, catalytic effects were studied in more detail by spiking possible catalysts to a surface water sample forming relatively high NDMA concentrations and low levels of NPYR in the standard chloramination test. The results of these experiments are shown in Table 3.9.

As it can be seen, formation of NDMA or NPYR during chloramination was in neither case substantially enhanced by the anions and cations tested. NDMA and NPYR formation rather turned out to be significantly lowered in the presence of thiocyanate and nitrite, implying that both anions might react faster with chloramine than precursor compounds relevant for nitrosamine formation.

Based on the findings presented in this section, nitrosamine formation from amines, at least from secondary amines, by chloramination has to be regarded as rather inefficient and neither matrix nor catalytic effects seem to improve conversion rates substantially. Based on a conversion rate of about 2.5%, the formation of 93 ng/L NDMA observed for the surface water in Table 3.8 would require a dimethylamine concentration of about 2.3 μg/L, the formation of 4.5 ng/L NPYR would be related to a concentration of about 0.13 μg/L pyrrolidine.

Table 3.8
Conversion of amines into the corresponding nitrosamines during chloramination

	Matrix	Primary amine concentration (ng/L)	Nitrosamine concentration after chloramination# (ng/L)	Conversion based on molar concentrations (%)
DMA → NDMA	drinking water	1,000	8.0	0.49
DMA → NDMA	drinking water	10,000	280	1.70
DMA → NDMA	surface water I	10,000	450	2.74
DMA → NDMA	surface water II	10,000	430	2.62
DEA → NDEA	drinking water	1,000	7.4	0.53
DPA → NDPA	drinking water	1,000	0.2	0.02
DBA → NDBA	drinking water	1,000	8.4	0.69
PIP → NPIP	drinking water	1,000	18	1.35
PYR → NPYR	drinking water	1,000	26	1.84
MOR → NMOR	drinking water	1,000	1.0	0.07

corrected for nitrosamine levels formed by the precursors being naturally present in the sample
Conditions: 0.4 mM chloramine, pH 7, 20 °C, 168 h

Table 3.9
Impact of possible catalysts on nitrosamine formation during chloramination

	NDMA concentration following chloramination (ng/L)	Change in NDMA-FP (%)	NPYR concentration following chloramination (ng/L)	Change in NPYR-FP (%)
Surface water	93	0	4.5	0
+ 0.5 mM $FeCl_2$	70	-25	3.6	-20
+ 0.5 mM $FeCl_3$	70	-25	3.3	-27
+ 0.5 mM $CuCl_2$	88	-5.4	3.5	-22
+ 0.5 mM $NiCl_2$	93	0	4.1	-8.9
+ 0.5 mM $NaNO_2$	29	-69	0.69	-85
+ 0.5 mM $NaNO_3$	93	0	4.8	+6.6
+ 0.5 mM NaCl	95	+2.2	4.5	0
+ 1.0 mM NaCl	92	-1.1	4.8	+6.6
+ 0.5 mM NaSCN	1.3	-99	<0.3	> -93

Conditions: 0.4 mM chloramine, pH 7, 20 °C, 168 h

CHAPTER 4
OCCURRENCE OF POSSIBLE PRECURSORS

LITERATURE REVIEW

According to current knowledge, there are always two key precursors involved in the two major pathways of nitrosamine formation: organic nitrogen compounds and an inorganic, nitrogen-containing species like dinitrogen trioxide (N_2O_3) or monochloramine (NH_2Cl). Relevant inorganic precursors have been described in detail in Chapter 3 "Formation of Nitrosamines" and include various species of the natural nitrogen cycle. Nitrite has been detected in rivers (up to 1 mg/L), in effluents of sewage treatment plants (more than 80 mg/L) and in industrial effluents (e.g. 500 mg/L at a galvanizing plant) (Oliver 1979a). Furthermore, nitrate occurs ubiquitously in the environment and microorganisms often contain enzymes which effectively reduce nitrate to nitrite.

Individual Organic Nitrogen Compounds

Organic nitrogen-containing compounds that might form the undesired nitrosamines are numerous. Among the diverse groups of amino compounds, secondary amines react rapidly with a variety of inorganic nitrogen-containing species to form N-nitroso compounds, while primary, tertiary and quaternary amines seem to react more slowly, according to more complex reaction mechanisms. Nitrosamine formation with tertiary nitrogen compounds for instance necessitates the breakage of the C-N bond and this process is often too slow to permit significant nitrosamine formation. This is particularly true for aliphatic tertiary amines like trimethylamine. However, slower reaction rates for tertiary amines in comparison to secondary amines can not be generalized. For the traditional nitrosation pathway, for instance, it is has been shown that nitrosamines are formed more rapidly from some tertiary nitrogen compounds (Figure 4.1) providing structural characteristics supporting the release of the nitrosated dialkylamine (Loeppky et al. 1994).

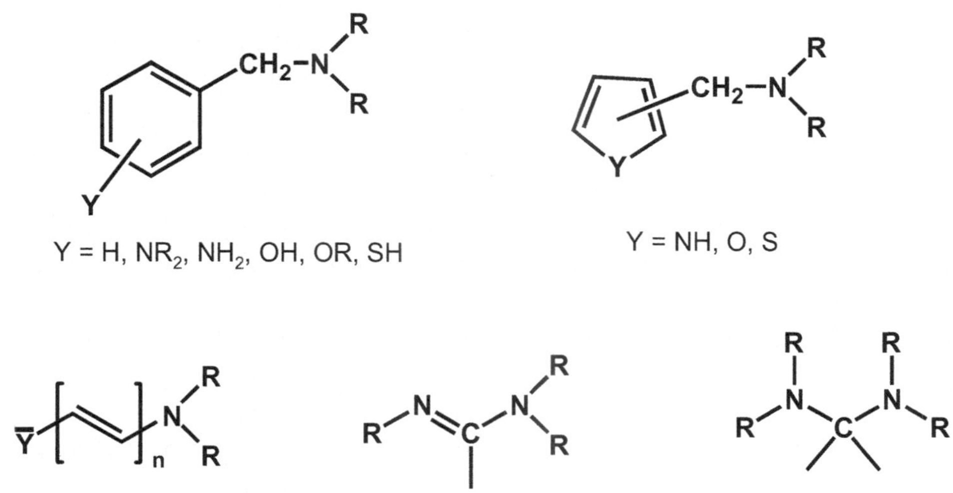

Figure 4.1 Molecular structures of tertiary amines favorable for nitrosamine formation

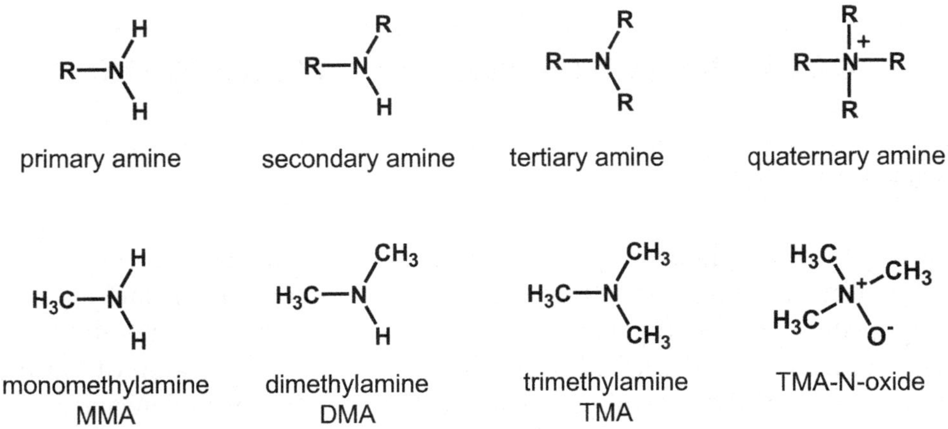

Figure 4.2 Molecular structures of aliphatic amines

Tertiary amines of this kind are for instance aminopyrine and gramine that nitrosate quite rapidly at room temperature to produce NDMA. Rapid nitrosamine formation from these precursors is made possible by assistance of an electron donating heteroatom, either attached to, or in the case of heterocyclic compounds, included in the aromatic ring. Electron donation from this moiety helps in releasing the nitrosamine. With regard to possible substituents, the following order of reactivity was observed: N >> S > O. Rapid nitrosamine formation was also shown for non-aromatic systems providing an electron donating heteroatom facilitating the release of the nitrosamine group by mesomerism (Loeppky et al. 1994).

Aliphatic Amines

Many aliphatic amines (Figure 4.2) are both important industrial chemicals and natural products of the metabolism of plants and microorganisms and thus are commonly found in aquatic environments. Dimethylamine (DMA) is the most frequently used amine among all aliphatic amines produced in the chemical industry. Worldwide production volumes of methylamines amount to 500,000 t/a. 57% of this amount is related to DMA (van Gysel and Musin 1997). Main application fields of DMA comprise the production of N,N-dimethylformamide (70%) and N,N-dimethylacetamide, the application as vulcanization accelerator in rubber industries (11%), and the production of fungicides and herbicides (10%, e.g. monuron, diuron, fenuron, fluometuron, chloroxuron, metribuzin, dicrotofos, diphenamid, oxamyl, ferbam, thiram). Furthermore, it is used as solvent, flotation chemical, absorber for acid gases, antioxidant, tanning agent, and insect attractant. DMA also represents an important intermediate in the production of other chemical products (dyes, chemicals for the textile and rubber industry, propellants, and pharmaceuticals) (Falbe and Regitz 1995; Landesumweltamt Brandenburg 2003). DMA is also a typical metabolic product of proteins in animals and plants. As a consequence, it is found in urine (Brooks et al. 1972), feces (van Rheenan 1962), higher plants (Smith 1971), and algae (Rolle et al. 1971). Investigations on diverse food stuffs (fish, bread, cheese, water, and fresh vegetables) revealed the highest DMA concentrations in beetroots (ca. 1 mg/kg) (Neurath et al. 1977).

Because DMA may be formed by demethylation of trimethylamine (TMA), this tertiary amine is also a possible NDMA precursor. TMA occurs in plants (Cromwell and Richardson 1966), fish, and algae (Ayanaba and Alexander 1974). Trimethylamine is produced during biodegradation of nitrogen-containing organic molecules in plants and mammals (e.g. amino acids, cell membrane structural lipid phosphatidyl choline (lecithin), trimethylamine-N-oxide), during putrefaction, and during dry distillation of sugar beet residues (precursor is betaine) (Simenhoff et al. 1976; Tricker et al. 1994; Falbe and Regitz 1995). Chemically fixed (mostly as choline), the trimethylamine functional group is found in numerous plants and fungi. Trimethylamine is also found in sea-water fish, the smell of herring brine is mainly caused by TMA (Falbe and Regitz 1995). Addition of trimethylamine to lake water or municipal sewage was shown to result in the microbial production of dimethylamine (Ayanaba and Alexander 1974). Following absorption from the gastrointestinal tract, trimethylamine is demethylated to dimethylamine and excreted via urine, gastric juice, or bile. Some trimethylamine is endogenously oxidized to trimethylamine-N-oxide. Typically, urine concentrations of the latter compound are twice as high as those of dimethylamine (Zuppi et al. 1997). Trimethylamine-N-oxide is also present in seafood, and may be broken down to trimethylamine by bacteria (Mitch et al. 2003a).

In chemical industries, diethylamine (DEA) is an important intermediate in the production of rubber, textile and flotation chemicals, synthetic resins, dyes, polymerization decelerators, insecticides, and in galvanic baths. The sulphamic salt of DEA is used as flame retardant (Falbe and Regitz 1995). Triethylamine (TEA) is used for the production of choline and choline salts, quaternary ammonium compounds, as (extraction) solvent, catalyst in various organic reactions, corrosion inhibitor, and as gas odorant (Falbe and Regitz 1995). Dipropylamine is an important intermediate in organic synthesis (Falbe and Regitz 1995). Dibutylamine (DBA) is used as intermediate in the production of dyes, emulsifying and flotation agents, specialty soaps, insecticides, anti-corrosion agents and as additive in vulcanization accelerators (Falbe and Regitz 1995). Diethanolamine (DELA) and triethanolamine (TELA) are used in the production of emulsions, cleaning products for furniture and floors, shoe creams, lubricants, pharmaceuticals, and cleansers. Furthermore DELA and TELA are used as absorbers for acid gases (H_2S and CO_2) and as important intermediates in the chemical synthesis of heterocyclic compounds (Falbe and Regitz 1995). Anthropogenic loads of alicyclic amines (e.g., morpholine, pyrrolidine, and piperidine) often result from the production of pharmaceuticals, dyes, pesticides, lacquers, optical brighteners, polymerization accelerators, and rubber. Pyrrolidine and piperidine might be also released from foodstuff (degradation product of the amino acid proline) and alkaloids, respectively.

Occurrence in the Environment

Due to the wide field of their technical and industrial application and partly due to their natural formation, these amines are found in various matrices relevant to the environment. All methylamines and ethylamines can be detected in human urine. The typical daily excretion amounts to 20-100 μg per person (Bittersohl and Heberer 1978). The average concentration of DMA in human urine is about 40 mg/L, human feces contains approximately 0.41 μg/g DMA (Tricker et al. 1994). Nitrosamine precursors were also found in manure (van Rheenan 1962; Bergstrom et al. 1972). Nitrate, nitrite and total amine nitrogen were found in amounts ranging from 50-200 mg/kg, 40-50 mg/kg, and 10-30 mg/kg, respectively (Bergstrom et al. 1972). Methyl-, dimethyl-, diethyl-, dipropyl-, dibutyl-, trimethyl-, triethyl-, and tripropylamine have been identified in soil (Golovnya et al. 1982). Dimethylamine concentrations in primary wastewater effluents are typically in the range of 20 to 80 μg/L (Mitch et al. 2003a, 2003b; Mitch and Sedlak 2004). Since dimethylamine is well degradable by bacteria, levels in secondary wastewater effluents are typically significantly lower (< 10 μg/L) (Islam et al. 1998; Mitch et al. 2003a, 2003b; Mitch and Sedlak 2004; Sedlak et al. 2005; Pehlivanoglu-Mantas and Sedlak 2006a), even though a few authors reported no significant loss of dimethylamine upon secondary treatment (Hwang et al. 1995; Abalos et al. 1999). Analyzing primary and secondary effluents of wastewater treatment plants revealed typical DMA removal efficiencies in the range of 30 to 95% during secondary treatment (Mitch and Sedlak 2004; Sedlak et al. 2005, Pehlivanoglu-Mantas and Sedlak 2006a). Studies on TMA concentrations in wastewater are rare, but were reported to be within the same range as DMA concentrations or lower (Pehlivanoglu-Mantas and Sedlak 2006a). Hwang et al. (1995) measured 35 µg/L TMA and 230 µg/L DMA in a wastewater effluent and Abalos et al. (1999) 300 µg/L TMA and 230 µg/L DMA in an effluent-dominated river. In industrial wastewater 75 µg/L DMA, 25 µg/L methylamine (MA) and 15 µ/L ethylamine (EA) were detected by Sacher et al. (1997). Also morpholine (3-20 µg/L) and piperidine (0.2-14 µg/L) were found in industrial and municipal wastewaters (Pietsch et al. 1996). Knepper et al. (1999) detected DMA at concentrations of 10-26 µg/L in industrial wastewaters of three chemical companies. Typical concentrations of individual amines (including DMA, DEA, morpholine, piperidine, and pyrrolidine) in German surface waters were in the range of < 0.1 to 5 µg/L (Pietsch et al. 1996, 1997, 1999; Sacher et al. 1997). For unpolluted surface waters dimethylamine concentrations of less than 0.1 µg/L were reported (Gerecke and Sedlak 2003; Mitch et al. 2003b).

The NDMA Precursor Test

Mitch et al. (2003b) developed a NDMA precursor test capable of quantifying the concentration of organic precursors of NDMA that could be formed during chloramination. The method involves applying a high dose of monochloramine to a pH-buffered sample followed by 10-day contact period. Starting with a solution of 73 µg/L DMA in deionized water, the application of this NDMA precursor revealed that about 0.5% DMA were converted to NDMA (ca. 569 ng/L) (Mitch et al. 2003b). Similar test results were obtained by Gerecke and Sedlak (2003) that spiked surface water with DMA and obtained a yield of about 0.6% NDMA. The test was then applied to waters from two wastewater treatment plant process units and two surface waters with known DMA concentrations (Mitch et al. 2003b). The results demonstrated that only 0.4-12% of the formed NDMA could be accounted for by DMA in wastewaters, and at maximum 3% by DMA in surface waters (on the basis of the results obtained for DMA in deionized waters). Gerecke and Sedlak (2003) estimated that the maximum percentage of NDMA precursors that could be explained by DMA in natural waters is 25%. Consequently, DMA does not seem to be the only precursor for NDMA formation in wastewaters and natural waters. Comparing the contribution of DMA to the overall precursor fraction for primary and secondary wastewater effluents revealed by similar estimations that DMA accounts for a significant percentage (ca. 70-90%) of the dissolved NDMA precursors in primary effluent, whereas it is not an important precursor (ca. 14%) in the dissolved fraction of the secondary effluent due to biodegradation (Mitch and Sedlak 2004, Sedlak et al. 2005). Application of the chloramination NDMA precursor test to trimethylamine, dimethylethanolamine, and dimethylaminobenzene in deionized water resulted also in appreciable NDMA concentrations, even though formed levels were slightly lower than those obtained for DMA (Mitch and Sedlak 2004). NDMA formation from methylamine, tetramethylammonia ion, methylethylamine, diethylamine, triethylamine and trimethyl-amine-N-oxide was substantially lower and is most probably not relevant for chloramination in the field (Mitch and Sedlak 2004).

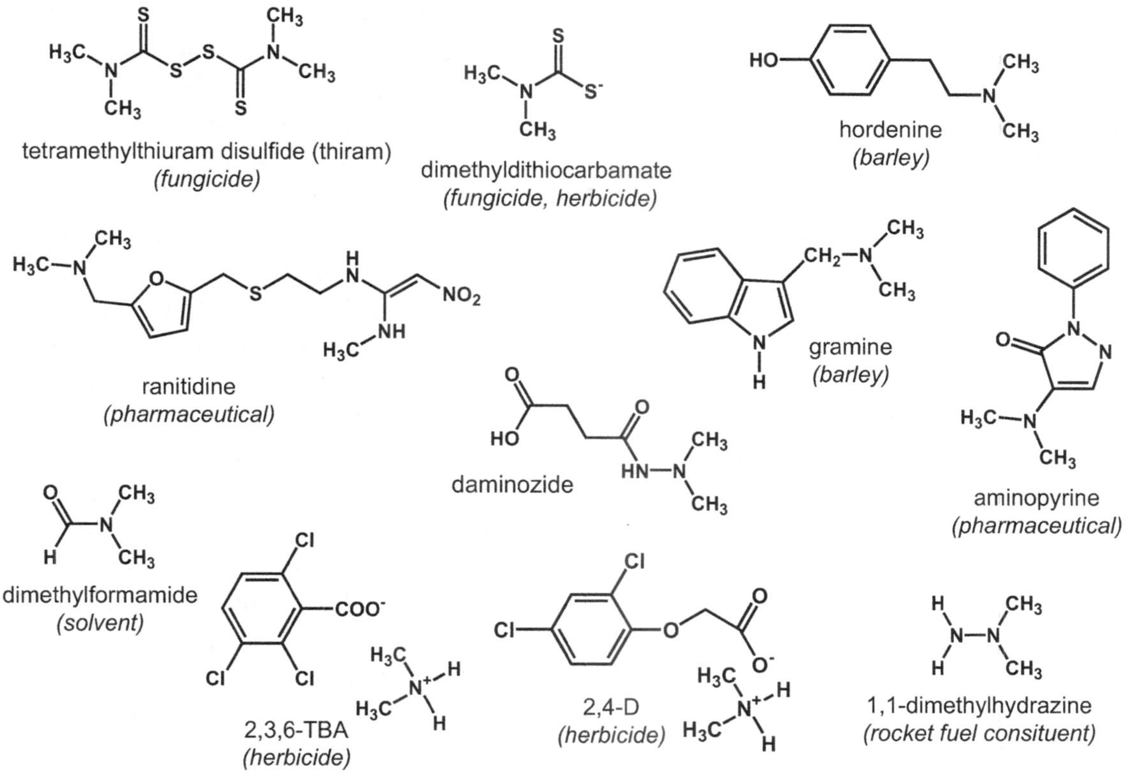

Figure 4.3 Molecular structures of important precursor compounds for NDMA

Other possible sources of amine precursors are found in foods, drugs, cosmetics, industrial and agricultural chemicals (solvents and pesticides), and tobacco (Preussmann 1984). In these fields, several nitrogen-containing organic compounds are found that provide dimethylamine functional groups that can directly or indirectly (via conversion into the corresponding secondary amines) result in the formation of nitrosamines (Figure 4.3). Dimethylformamide that is often used as aprotic solvent in chemical industries can form dimethylamine in water. This process is catalyzed in acidic or alkaline media (Falbe and Regitz 1995). Furthermore, dimethylformamide (as well as dimethylcyanamide) is a chloramination by-product. Application of the NDMA precursor test revealed that dimethylformamide might be a relevant NDMA precursor, even though formed NDMA levels were about an order of magnitude lower when compared to DMA (Mitch and Sedlak 2004). Testing dimethylcyanamide caused NDMA levels that were about three orders of magnitude lower than those formed from DMA (Mitch and Sedlak 2004).

Furthermore, several pesticides may serve as a source of amines. The introduction of dimethyl- and diethyldithiocarbamates into soil resulted in the accumulation of secondary amines in amounts of up to 0.25% of the pesticide mass applied (Tate and Alexander 1974). Furthermore, dimethyldithiocarbamate is often contaminated with NDMA and can, after application, contribute to pulses of NDMA in raw sewage (Mitch and Sedlak 2004). In the chloramination NDMA precursor test, dimethyldithiocarbamate formed substantial NDMA amounts that were comparable to those of DMA and TMA (Mitch and Sedlak 2004). Formation of NDMA with dimethylamine as intermediate is also possible from the fungicide thiram (tetramethylthiuram disulfide) introduced into soil (Ayanaba et al. 1973; Ayanaba and Alexander 1974; Najm and Trussell 2001). It was found that the reaction of thiram with chlorine (3 mg/L) even yielded higher amounts of NDMA than the corresponding reaction with DMA, even though NDMA yields were generally low, with only 0.005% for DMA and 0.2% for thiram (Najm and Trussell 2001). Diaminozide, the active component in the plant growth regulator AlarR, can be hydrolyzed to unsymmetrical dimethylhydrazine (UDMH) and then be oxidized to NDMA (Sen et al. 1975; Pylypiw et al. 1992). Furthermore, NDMA has been a contaminant of several herbicides formulated as dimethylamine salts in concentrations ranging from 0.3 to 640 mg/kg (Oliver 1979a). Representatives of this class are 2,4-D and 2,3,6-TBA (Oliver 1979a; Child et al. 1996; Mitch et al. 2003a). It was postulated that nitrite, used as a corrosion inhibitor in metal containers, reacted with dimethylamine during storage (Oliver 1979a). However, it is also possible that NDMA will be formed in the environment in the course of their application. The herbicide trifluraline can act as a source of NDPA (Oliver 1979a). Dimethyldithiocarbamate and thiram are also used as accelerators in certain vulcanization processes of the rubber industry.

Several nitrosamines have been identified in personal care products over the last 25 years. These compounds are primarily formed from amine precursors and nitrosating agents in the cosmetic product but are also introduced into the product through cosmetic raw materials. Nitrosodiethanolamine (NDELA) is the nitrosamine found most frequently in cosmetic products. Market surveys of sunscreens and cosmetics have revealed NDELA levels of up to 45 mg/kg (Fan et al. 1977; Havery and Chou 1994). In cosmetics, soaps and shampoos, the major source for NDELA is triethanolamine, an antiallergenic agent (Brunnemann et al. 1982). Furthermore, di- and triethanolamine and some of their salts and amides are used on a large scale as emulsifiers in many cosmetic formulations (Preussmann 1984).

In addition to this, pharmaceutical products containing dialkylamine functional groups might serve as nitrosamine precursors. Aminophenazone (aminopyrine) has been taken of the markets due to high NDMA levels (ca. 1-370 μg/kg) detected in the formulated drug (Eisenbrand et al. 1979; Lijinsky et al. 1982). Low amounts of nitrosamines have also been found in other drugs (Krull et al. 1980). Another relevant pharmaceutical containing a dimethylamine moiety that can act as NDMA precursor is ranitidine (Mitch et al. 2003b).

1,1-Dimethylhydrazine has been used as a rocket propellant (Aerozin-50) and as intermediate in the production of photo and agricultural chemicals, pharmaceuticals and dyes. It is also used as gas absorbent for CO_2 and SO_2 (Falbe and Regitz 1995). UDMH can be released to the environment during its production or use. One of the primary ways 1,1-dimethylhydrazine enters the environment is from its use as rocket fuels. Some environmental contamination might be also caused by accidental spills and leaks from storage and waste sites. Small amounts of UDMH have been found in tobacco products (Choudhary and Hansen 1998). Furthermore, monomethylhydrazine and UDMH are products of the reductive metabolism of NDMA in mammals (Grilli and Prodi 1975; Godoy et al. 1984). UDMH can be oxidized in aqueous systems to NDMA. Degradation of UDMH with dissolved oxygen or hydrogen peroxide is catalyzed by metal ions, particularly copper. The reaction rate is strongly influenced by pH and fastest reactions were observed in alkaline solutions (Banerjee et al. 1984; Lunn and Sansone 1994; Choudhary and Hansen 1998).

The occurrence of NDMA in malt and beer has received considerable attention (Wainright 1986; Slack 1980). Concentrations of various nitrosable amines in barley and malt are not constant, but all are present in sufficient amounts to form some amount of NDMA, particularly during kilning. DMA occurs in malt and barley in mg/kg quantities and is presumably a normal product of barley metabolism. Further important NDMA precursors occurring in barley at mg/kg levels are natural tertiary amines like gramine and hordenine (Slack 1980; Wainright 1986). Complete absence of nitrogen oxides during kilning results in the virtual absence of NDMA from malt and beer. Following major investments to install either so-called low NO_x burners or indirectly heated kilns, NDMA levels formed during malting could be considerably lowered (Wainright 1986).

Natural Organic Matter and Dissolved Organic Carbon and Nitrogen Compounds

A major part of nitrosamine formation during chloramination seems to be related to either unknown anthropogenic or naturally occurring nitrogen-containing organic molecules in waters (organic nitrogen). Natural organic matter (NOM) has been demonstrated as possible NDMA precursor in studies by Chen and Valentine (2005), Luo et al. (2005), and Westerhoff et al. (2006).

Natural Organic Matter

Leaching from soils, diffusion from sediments, plankton and bacteria can be considered as the main sources of natural organic matter (NOM) in surface waters. Natural organic matter consists of humic substances – humic and fulvic acids – and non-humic matter, mainly proteins and amino acids, sugars and polysaccharides. The heterogeneity of humic substances derives not only from size differences between molecules but also from differences in chemical structure. In most natural waters, about 50% of the dissolved organic carbon (DOC) is composed of humic substances, the molecular weight of which ranges from 500 to 10,000 Da. There is evidence that the molecules form aggregates as their concentration increases (Nikolaou and Lekkas 2001).

Typically 90% of the dissolved humic substances in natural waters consist of fulvic acids and 10% of humic acids, a composition being contrary to that of humic substances from soils, where humic acids are in very large excess over fulvic acids. Reservoir waters have been found to be richer in DOC than river waters (Nikolaou and Lekkas 2001). It has been estimated that aquatic humic substances have a moderate aromatic character (ca. 25% of the total carbon content) and contain mainly carboxyl groups, phenolic groups, alcohol OH groups, methoxyl groups, ketones, and aldehydes. H:C atomic ratios indicate abundance of aliphatic functional groups (fatty acids, $-CH_2-$, $-CH_2CH_3$, $-CH_3$). Possible sources of aliphatic compounds include plant waxes from vascular plants and long-chain cell-wall lipids from algae, or the degradation of aquatic humic substances by the elimination of oxygen-containing functional groups (alcohols and ethers). In NOM molecules, chromophores (i.e. functional groups containing electrons that are promoted when light absorption occurs) that absorb in the UV region (< 400 nm) are substituted aromatic groups, including mono- and poly-substituted phenols and various aromatic acids. These chromophores belong mainly to the humic fraction of NOM. Characterization of NOM in regard to its functional groups is very important in order to determine the behavior of NOM during water treatment. Coagulation of fulvic acids, with a higher carboxylic and phenolic acidity than humic acids, is more difficult than coagulation of humic acids. The same is true for adsorption on granular activated carbon. The solubility of humic and fulvic acids is also influenced by functional groups (Nikolaou and Lekkas 2001).

Organic Nitrogen

Organic nitrogen compounds comprise a relatively small portion (0.5 to 10% by weight) of the total NOM mass (Westerhoff and Mash 2002) and a variety of organic compounds, many of which come from diverse natural and anthropogenic sources. The total organic nitrogen (TON) is defined as the sum of particulate organic nitrogen (PON) and dissolved organic nitrogen (DON). The carbon to nitrogen ratio for NOM in drinking water ranges between 10 and 30 mg C/mg N (Westerhoff et al. 2006).

Nitrogenous organic matter consists of aliphatic or cyclic structures. Free amino acids represent the major pool of low molecular weight organic nitrogen-containing matter, and are categorized either as acids, bases or neutrals. Amino acids incorporated into larger NOM molecules (i.e. combined or hydrolysable amino acids), heterocyclic-N, polyamine peptides and biomolecular material (i.e. RNA, DNA) represent the dominant pool of larger molecular weight organic nitrogen-containing compounds. The nitrogen atom in the nitrogenous matter has a lone pair of electrons and acts as a nucleophile, which is important for its chemical and biochemical reactivity (Westerhoff and Mash 2002).

The precise structural composition of organic nitrogen-containing matter is still under discussion, primarily due to analytical limitations and its distribution among a wide range of molecular weights (Westerhoff and Mash 2002; Pehlivanoglu-Mantas and Sedlak 2006a). Amino acids appear to constitute between 30% and 50% of the DON, with less than 5% present as free amino acids. In natural systems glutamic acid, glycine, serine, and aspartic acid are the dominant amino acids, while in wastewater- or manure fertilizer-impacted systems, aromatic amino acids may also be present. Cell walls (peptidoglycans) or membranes (lipopolysaccharides) and recondensation products (melanoidin-like) are amides and constitute approximately 20% of the DON. Heterocyclic-N compounds (e.g. pyroles, pyrrolidines, pyridines, pyrines and pyrazoles) may account for 25% of the organic nitrogen-containing compounds. Specific compounds within amide and heterocyclic-N or the remaining organic nitrogen-containing material (e.g. nitriles, nitrosamines) have not been well characterized. Basic and neutral NOM fractions are N-enriched compared to acid fractions, and transphilic and colloidal NOM fractions (DOC/DON of 5-10 mgC/mgN) are N-enriched relative to hydrophobic acid fractions. Colloidal NOM fractions consist of decomposed biological cellular material and are most enriched with aminosugars (Westerhoff and Mash 2002; Westerhoff et al. 2006). The colloidal and other nitrogen enriched NOM fractions were shown to have a greater NDMA formation potential during chloramination than the hydrophobic fraction (Luo et al. 2005; Westerhoff et al. 2006).

Operationally, components of the DON pool have been divided into high molecular weight (HMW, usually > 1 kDa) and low molecular weight (LMW) compounds. HMW DON includes proteins (such as enzymes, modified bacterial wall proteins, dissolved combined amino acids, nucleic acids (DNA, RNA) and humic-like substances that have a relatively low nitrogen content. Things are even more complicated due to the fact that some LMW and HMW DON compounds may be loosely held or adsorbed to humic substances. Organic nitrogen compounds that make up the LMW DON pool include urea, peptides, dissolved free amino acids, amino sugars, purines, pyrimidines, pteridines, amides, methyl amides and others (Berman and Bronk 2003). In finished drinking waters, approximately half of the DON consists of LMW compounds (Westerhoff et al. 2006).

In natural waters, the DON constitutes about 50% of the TON. Average DON concentrations for surface waters were 0.3 mg N/L, but ranged by three orders of magnitude from < 0.01 to 10 mg N/L (Westerhoff and Mash 2002). During periods of warm water temperatures and algal activity, DON concentrations tend to increase (Westerhoff et al. 2006). Discharging wastewater into surface water increases DON concentrations, since wastewater contains higher DON levels than non-impacted surface waters (Westerhoff et al. 2006). DON concentrations in secondary treated wastewater effluent typically range from 1 to 5 mg N/L (Pehlivanoglu-Mantas and Sedlak 2006a). Mean DON concentrations in raw and finished waters from drinking water treatment plants in the United States were 0.19 mg N/L and 0.15 mg N/L, respectively, with DON concentrations ranging from < 0.05 to 0.45 mg N/L (Westerhoff et al. 2006). Typically, DON concentrations in groundwaters are lower than in surface waters (Westerhoff and Mash 2002).

Primary sources of DON include agricultural fertilizers (manure, urea), wastewater discharges, and eutrophic bodies of water (Westerhoff and Mash 2002). Site-specific variation of DON concentrations in lakes was associated with seasonal eutrophication, and in rivers with urban runoff events or stream discharge. Algae produce protein-rich material (i.e. DON) and may cause seasonal or rapid changes in DON concentrations in drinking water supplies. The eutrophication status of a reservoir affects DON levels. Bacteria readily biodegrade proteinaceous and carbohydrate material released by algae, but in turn release soluble microbial products that can become humified. Spatial and temporal fluxes and concentrations of the DON follow those of the DOC. DOC/DON ratios range from < 10 in systems with high algae productivity, wastewater discharge impacts, or agricultural activity to > 40 from forested watersheds or those with well developed soils in temperate climate zones (Westerhoff and Mash 2002).

Organic Carbon

It was found by Gerecke and Sedlak (2003) that NDMA precursors in surface waters exhibit a correlation with dissolved organic carbon. However, this correlation was not always found (Najm and Trussell 2001) and is thus not very reliable to predict the NDMA formation potential of waters. Mitch and Sedlak (2004) investigated several individual compounds and organisms that might form NDMA during chloramination in their NDMA precursor test. According to these experiments neither important DOC constituents of wastewater effluents like hippurate, creatinine, albumin, histidine, N-acteylglucosamine nor pure cultures of microorganisms such as Gordna amarae, Escherichia coli and Saccharomyces cerevisiae or samples of lysed cells of Escherichia coli were of relevance for NDMA formation during chloramination (Mitch and Sedlak 2004). Some NDMA formation was observed with phosphatidylcholine, however, produced amounts were about three orders of magnitude lower than those obtained from DMA (Mitch and Sedlak 2004).

Size separation experiments by Mitch and Sedlak (2004) revealed the existence of two types of precursors in secondary wastewater effluents: precursors associated with particles larger than 0.7 µm and small, dissolved precursors of less than 3000 Da. Significant concentrations of particle-associated NDMA precursors were only detected in secondary effluents of treatment plants that recycled water from sludge thickening operations in which dimethylamine-based polyelectrolytes were used (Mitch and Sedlak 2004). Gerecke and Sedlak (2003) found that nearly no NDMA precursors were associated with particles greater than 0.7 µm in surface waters.

Although secondary biological treatment is capable of effectively removing the important NDMA precursor dimethylamine and about 90% of the DON (Pehlivanoglu-Mantas and Sedlak 2006a), it turned out to be rather ineffective in removing other dissolved NDMA precursors, since the average percentage reduction during secondary treatment in several plants was only about 70% (Mitch and Sedlak 2004; Sedlak et al. 2005). Mitch and Sedlak (2004) concluded that dissolved precursors other than dimethylamine resist biodegradation after demonstrating that NDMA precursors are not created by microbial DON degradation. Further experiments demonstrated that the dissolved NDMA precursors in secondary wastewater effluents can not be removed by microfiltration, but by reverse osmosis treatment, pointing to charged molecules like secondary or tertiary amines. Thus, uncharged, low molecular weight dimethylamides do most likely have no major impact on NDMA formation during wastewater chlorination (Mitch and Sedlak 2004).

A series of experiments was conducted by Gerecke and Sedlak (2003) to identify NDMA precursors in natural water samples collected from reservoirs, lakes, groundwaters, and in natural organic matter isolates. According to direct analytical measurements, DMA can not account for the major part of NDMA precursors in surface water. Based on sorption studies using different SPE materials, the authors hypothesized that NDMA precursors in surface water were associated with humic substances and other high molecular weight polymers (Gerecke and Sedlak 2003). However, the authors could not deduce any correlation between the DON or NOM and the NDMA formation potential, even though the NDMA precursor potential was somehow related to the DOC (Gerecke and Sedlak 2003). On the other hand, sampling of two reservoirs over a period of 7 months at different depths revealed that the concentration of NDMA precursors was significantly higher in the epilimnion than in the hypolimnion during the period of thermal stratification despite similar concentrations of DOC. After turnover, the concentrations of precursors were similar throughout the water column (Gerecke and Sedlak 2003). Chloramination of surface waters in laboratory-based experiments under conditions that are comparable to those found in drinking water treatment indicated that the concentration of NDMA produced from naturally occurring precursors usually will be less than 0.1 nM (Gerecke and Sedlak 2003).

Polyelectrolytes and Ion-Exchange Resins

Natural organic matter poses a problem in water treatment for a number of reasons. Apart from the formation of nitrosamines and other disinfection by-products, NOM might cause taste and odor problems and it contributes to the fouling of membranes and ion-exchange resins, the deterioration of water quality in distribution systems because of bacterial growth, and the more difficult oxidation of dissolved iron and manganese to insoluble forms (Bolto et al. 2002). Typical NOM concentrations in drinking water sources are in the range of 2 to 10 mg/L, although much higher levels are sometimes present (Bolto et al. 2002). Depending on the nature of the NOM, coagulation with cationic polymers or alum and ion exchange are among the treatment steps used for NOM removal in several water treatment plants (Bolto et al. 2002).

Polyelectrolytes

Polyelectrolytes used in water treatment as primary coagulants or as floc builders, however, are often an important source of NDMA precursors during drinking water treatment (Kohut and Andrews 2002, 2003a; Wilczak et al. 2003). Depending on their charge, polyelectrolytes can be classified as cationic, anionic or non-ionic polymers. Since most particulates such as clays and bacteria are negatively charged, cationic polymers are most commonly used for their additional ability to neutralize particulate charge, facilitating agglomeration (Kohut and Andrews 2002).

Poyldiallyldimethylammonium chloride (polyDADMAC) is the cationic polyelectrolyte most commonly used as primary coagulant in drinking water treatment (Kohut and Andrews 2003a). PolyDADMAC polymers are manufactured using DMA (which can react with allyl chloride to form diallyldimethylammonium chloride monomers, which are then polymerized) and the polymers' internal structure contains DMA-like molecule fragments (Wilczak et al. 2003; Najm et al. 2004). Commercial polyDADMAC polymer solutions may contain residual DMA left over from the manufacturing process and DMA concentrations ranged from 0.21 µg/mg of active polymer ingredient to 1.2 µg/mg (or 0.02% to 0.12% by weight) (Najm et al. 2004). Other cationic polyelectrolytes are epichlorhydrin-dimethylamine (epi-DMA), dimethylamine-based polyacrylamide and N,N,N-trimethyl-2[(1-oxo-2-propenyl)oxy]ethan-aminium-based polyacrylate (polyADAMQUAT), also providing the relevant dimethylamine group in each polymer structure (Kohut and Andrews 2003a; Mitch and Sedlak 2004) (Figure 4.4).

It was shown in several laboratory-based and field studies that the use of polyDADMAC and epi-DMA in the coagulation/flocculation/sedimentation process in conjunction with chlorination causes significant NDMA formation (Child et al. 1996; Najm and Trussell 2001; Kohut and Andrews 2002, 2003a; Wilczak et al., 2003; Najm et al. 2004). Epi-DMA resulted in even higher NDMA yields than polyDADMAC (Kohut and Andrews 2002, 2003a). Most interestingly, within the chloramination NDMA precursor test, the dimethylamine-based polyacrylamide polymer produced approximately 400 times more NDMA per mol nitrogen than DMA, demonstrating its importance as precursor in water treatment plants (Mitch and Sedlak 2004). On the other hand, the polyelectrolyte polyADAMQUAT did not form significant NDMA concentrations during the same test with levels about four orders of magnitude lower than those obtained for the DMA-based polyacrylamide (Mitch and Sedlak 2004). Non-ionic polymers based on polyacrylamide, polyacrylate anionic polymer, and alum most probably also do not contribute to NDMA formation (Child et al. 1996; Kohut and Andrews 2002; Wilczak et al. 2003; Mitch and Sedlak 2004; Najm et al. 2004). For polyDADMAC and epi-DMA, NDMA levels increased with polymer and chloramine concentrations (Kohut and Andrews 2003b; Wilczak et al. 2003; Najm et al. 2004) and longer chloramine contact times (Wilczak et al. 2003; Mitch and Sedlak 2004; Najm et al. 2004).

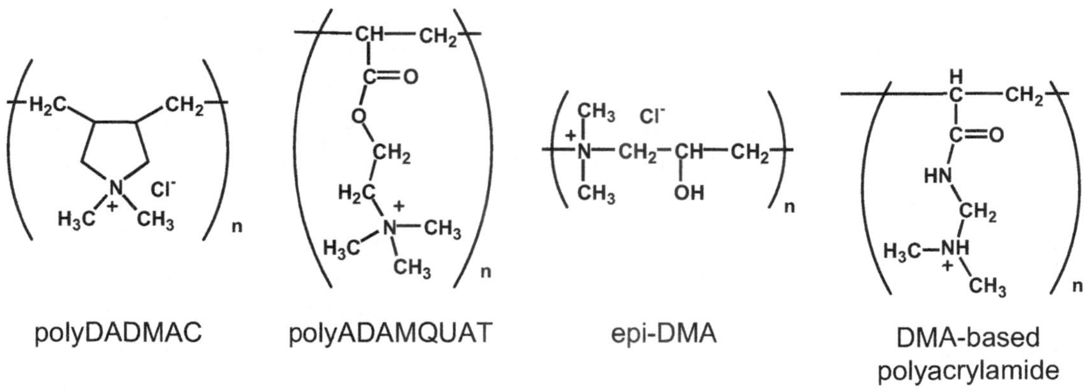

Figure 4.4 Structures of polyelectrolytes used in water treatment

Furthermore, NDMA formation from polyDADMAC polymers turned out to be pH-dependent. Within the tested range of pH 7.0-9.5, NDMA formation showed a maximum at pH 8 (Najm et al. 2004). When exposed to chloramine, polyDADMAC polymers release DMA into the water as a result of the breakdown of the polymer itself in addition to their intrinsic DMA residues (Najm et al. 2004). The amount of DMA released increases with increasing polymer and chloramine concentrations (Najm et al. 2004). However, released DMA is not necessarily the sole precursor of NDMA formation and it is rather probable that the added chloramine can also react directly with the polymer molecules to form NDMA (Najm et al. 2004). Investigations by Wilczak et al. (2003) demonstrated that several polyDADMAC polymers sold by different suppliers under different commercial names formed similar concentrations of NDMA during chloramination in the lab.

It was also shown that NDMA might be formed in processes with free chlorine disinfection, even though chlorine concentrations necessary for NDMA formation in this case exceeded the concentrations normally used in water treatment (Child et al. 1996; Najm and Trussell 2001). Decreasing the polymer doses in two treatment plants resulted in a significant reduction of NDMA concentrations in the drinking water (Wilczak et al. 2003). Similarly, NDMA concentrations were significantly lower in drinking water after the use of amine-based polymers was terminated in some treatment plants in Canada (Andrews and Taguchi 2000; Najm and Trussell 2001).

Ion-Exchange Resins

Anion exchange resins are commonly used in various ion-exchange processes in drinking water treatment and consist primarily of a polymer matrix and an ion-active group. In common strong and weak anion exchange resins, the ion-active group is often a quaternary amine function (Figure 4.5) (Kimoto et al. 1980).

Several studies have reported on detectable NDMA levels in waters exposed to certain anion exchange resins and it was concluded that most likely this NDMA originated from quaternary amines used to prepare these resins (Fiddler et al. 1977; Gough et al. 1977; Angeles et al. 1978; Cohen and Bachman 1978; Kimoto et al. 1980; Loveland et al. 2001; Najm and Trussell 2001; Wilczak et al. 2003).

Apart from NDMA, Fiddler et al. (1977) found NDEA in water exposed to deionizing resins. NDMA concentrations were in the range of 0.03-0.34 μg/L, those for NDEA in the range of 0.33-0.83 μg/L. The authors mentioned that the highest nitrosamine concentrations were found after resin regeneration and that at elevated temperatures (78 °C) trimethylamine was released from the resin (Fiddler et al. 1977). The origin of the nitrosamines, however, is not clear. There might be either impurities caused by the manufacturing process (Gough et al. 1977) or nitrosamines are formed during application of the resin to certain types of water, eventually promoted via surface-catalyzed reactions (Angeles et al. 1978; Mitch et al. 2003a). Gough et al. (1977) reported on a NDMA contamination of a resin with concentrations of up to 125 μg NDMA per kg resin. Angeles et al. (1978) suggested that mixed-bed resins promote nitrosation of precursors because proton displacement on cationic resins creates acidic sites favorable for nitrosation of amine precursors as part of the anionic resin in the direct neighborhood.

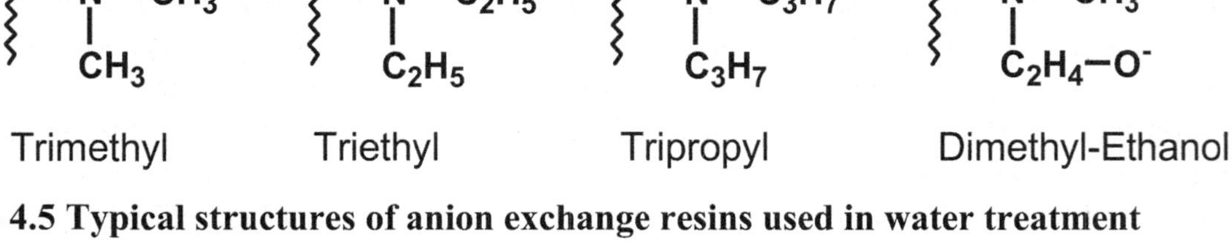

Figure 4.5 Typical structures of anion exchange resins used in water treatment

One study showed high levels of NDMA (1000-2000 ng/L) when tap water with 0.6-1.5 mg/L free chlorine was exposed to a mixed-bed demineralization resin (Kimoto et al. 1980). Following deactivation of the chlorine via passage of the tap water over a granular activated carbon (GAC) filter or by gently boiling the water prior to resin application, NDMA formation dramatically decreased (Kimoto et al. 1980; Najm and Trussell 2001). Najm and Trussell (2001) exposed four strong-base anion exchange resins with four different functional groups to groundwater, deionized water and deionized water that was spiked with 1 mg/L nitrite with no further treatment. None of the test waters contained any chlorine residual. Their results indicated that NDMA can be formed (ca. 50 ng/L) upon exposure of some strong-base anion exchange resins to both natural and organic-free waters without any chlorine background. Of the various resins tested only those containing the dimethylamine structural elements (dimethyl-ethanol, trimethyl) resulted in NDMA formation. However, the others might have promoted the formation of NDEA and NDPA, which was not further investigated. Most interestingly, the resin with the dimethylethanol quaternary amine functional group appeared to cause higher NDMA levels than the trimethylamine resin. The addition of nitrite in distilled water caused a substantial increase of NDMA formation during chloramination experiments with ion-exchange resins (Najm and Trussell 2001).

NATURE AND OCCURRENCE OF NITROSAMINE PRECURSORS

Investigations described in this section were performed in order to gain further insight into the nature and origin of precursors responsible for NDMA formation during chloramination. The studies should also clarify to what extent other nitrosamines than NDMA are relevant for water suppliers utilizing chloramination. For this, about 100 samples of different waters (i.e. different sampling locations and different sampling dates) and water types (river waters, lake waters, spring waters, and groundwaters) were chloraminated with the NFP test and formed nitrosamines were determined by chemical analysis. Additionally, tested water samples were characterized by a set of physical-chemical parameters. Tables 4.1 and 4.2 give a compilation of all analytical results obtained from these experiments. For characterizing the water samples, a classification system was used. The characters A, B, C denote different water systems, e.g. different rivers or different groundwater bodies. The Roman numerals (I, II, III) denote different sampling locations at one river. In Table 4.1, a rating of the wastewater load for each of the investigated waters is added based on the authors' experience.

Table 4.1
Characteristics of waters under investigation

	Water type (sampling date)	*pH*	*TOC (mg/L)*	*UV_{254nm} (m^{-1})*	*Alkalinity (mg/L $CaCO_3$)*	*NH_4^+ (mg/L)*	*NO_2^- (mg/L)*	*NO_3^- (mg/L)*	*Boron (mg/L)*	*Waste water load*
A	***river A, I*** *(31.01.2005)*	8.07	1.6	3.8	147	0.08	0.05	7.4	0.030	medium
B	***river A, II*** *(10.09.2004)*	8.04	1.6	3.8	117	0.01	0.04	5.1	n.d.	medium
C	***river A, II*** *(08.10.2004)*	n.d.	n.d.	n.d.	n.d.	n.d.	n.d.	n.d.	n.d.	medium
D	***river A, II*** *(22.10.2004)*	7.94	3.0	6.8	124	0.05	0.08	6.9	0.041	medium
E	***river A, II*** *(01.11.2004)*	8.01	3.0	7.6	129	0.05	0.07	8.1	0.039	medium
F	***river A, II*** *(31.01.2005)*	7.85	2.1	5.0	137	<0.01	<0.01	9.9	0.031	medium
G	***river A, II*** *(07.06.2005)*	n.d.	n.d.	n.d.	n.d.	n.d.	n.d.	n.d.	n.d.	medium
H	***river A, II*** *(29.06.2005)*	8.08	2.0	4.0	127	0.02	0.07	6.0	0.032	medium
I	***river A, III*** *(30.10.2004)*	8.12	2.8	7.0	127	0.06	0.06	7.7	0.040	medium
J	***river A, IV*** *(30.10.2004)*	7.94	3.3	7.9	129	0.16	0.05	7.9	0.040	medium-high
K	***river A, V*** *(30.10.2004)*	7.98	3.3	8.0	142	0.15	0.05	10.1	0.070	medium-high
L	***river A, VI*** *(04.07.2005)*	8.10	2.4	5.1	134	0.03	0.08	8.3	0.053	medium
M	***river A, VII*** *(08.11.2004)*	8.07	2.8	7.0	137	0.02	0.02	10.0	0.050	medium
N	***river A, VII*** *(04.07.2005)*	7.95	2.4	5.6	135	0.07	0.03	9.3	0.060	medium
O	***river A, VIII*** *(22.10.2004)*	n.d.	n.d.	n.d.	n.d.	n.d.	n.d.	n.d.	n.d.	medium
P	***river A, IX*** *(04.07.2005)*	8.17	2.6	5.5	132	0.04	0.02	4.6	0.060	medium-high
Q	***river A, X*** *(09.09.2004)*	8.14	1.9	5.2	125	0.01	<0.01	8.2	n.d.	medium-high
R	***river A, X*** *(04.11.2004)*	8.02	3.0	8.2	133	0.03	0.02	11.1	0.058	medium-high
S	***river A, X*** *(17.01.2005)*	7.98	2.8	6.7	137	0.03	0.08	16.1	0.061	medium-high
T	***river A, X*** *(30.01.2005)*	7.79	3.0	8.4	121	<0.01	0.01	16.0	0.045	medium-high
U	***river A, X*** *(04.07.2005)*	8.06	2.5	5.7	128	0.04	0.02	9.5	0.063	medium-high
V	***river A, X*** *(11.07.2005)*	8.03	2.2	5.5	130	0.03	0.01	8.8	0.064	medium-high
W	***river A, XI*** *(04.11.2004)*	7.93	3.1	8.5	133	0.03	0.03	11.0	0.058	medium-high
X	***river A, XI*** *(04.07.2005)*	8.02	2.7	6.0	130	0.04	0.03	9.5	0.068	medium-high

(continued)

Table 4.1 (Continued)

	Water type (sampling date)	pH	TOC (mg/L)	UV_{254nm} (m^{-1})	Alkalinity (mg/L $CaCO_3$)	NH_4^+ (mg/L)	NO_2^- (mg/L)	NO_3^- (mg/L)	Boron (mg/L)	Waste water load
Y	river A, XII (06.11.2004)	8.02	3.5	8.2	137	0.03	0.03	11.1	0.070	medium-high
Z	river A, XII (04.07.2005)	7.99	2.9	6.5	131	0.05	0.04	10.3	0.077	medium-high
AA	river A, XIII (06.11.2004)	8.04	3.2	8.2	137	0.04	0.03	11.2	0.069	medium-high
AB	river A, XIV (06.11.2004)	8.09	3.0	8.2	135	0.03	0.01	11.1	0.061	medium-high
AC	river B, I (26.10.2004)	8.13	3.7	9.9	228	0.10	0.10	18.3	0.050	medium-high
AD	river B, II (26.10.2004)	8.20	4.0	10.1	233	0.15	0.12	18.8	0.150	high
AE	river B,III (04.07.2005)	7.88	4.1	9.2	183	0.11	0.04	17.5	0.115	high
AF	river C (26.10.2004)	8.17	3.2	9.8	217	0.04	0.08	11.6	0.051	medium
AG	river D (06.11.2004)	7.41	8.9	19.2	256	0.16	0.17	17.1	0.580	very high
AH	river D (30.01.2005)	7.31	9.5	17.8	287	6.70	0.42	10.0	0.440	very high
AI	river D (04.07.2005)	7.45	8.1	18.0	204	0.26	0.26	15.3	0.378	very high
AJ	river E (06.11.2004)	8.09	4.0	10.2	243	0.09	0.14	22.9	0.300	very high
AK	river E (30.01.2005)	7.93	4.5	11.5	235	0.28	0.32	29.9	0.150	very high
AL	river E (04.07.2005)	8.08	4.1	9.8	203	0.08	0.11	20.7	0.236	very high
AM	river F, I (03.07.2005)	7.81	3.1	8.2	94	0.14	0.31	13.9	0.056	low-medium
AN	river F , II (04.11.2004)	7.71	2.6	5.6	108	0.05	0.08	12.9	0.089	medium
AO	river F , III (30.01.2005)	7.64	1.7	4.7	74	0.20	0.11	15.5	0.060	medium-high
AP	river G (04.11.2004)	7.95	3.9	9.8	141	0.17	0.15	16.0	0.080	medium-high
AQ	river G (30.01.2005)	7.73	3.9	11.4	117	0.18	0.15	19.2	0.041	medium-high
AR	river H (06.09.2004)	n.d.	3.4	9.6	180	0.02	0.05	16.8	0.100	medium-high
AS	river H (29.11.2004)	7.93	5.7	27.5	132	0.10	0.14	20.0	0.061	high
AU	river I (25.11.2004)	7.21	6.2	27.5	16	0.01	<0.01	3.4	<0.01	low
AV	river J (25.11.2004)	7.93	9.7	43.7	22	<0.01	<0.01	3.0	<0.01	low
AW	river K (08.11.2004)	7.94	3.2	6.9	86	0.01	0.02	17.9	0.100	very high

(continued)

Table 4.1 (Continued)

	Water type (sampling date)	pH	TOC (mg/L)	UV_{254nm} (m^{-1})	Alkalinity (mg/L $CaCO_3$)	NH_4^+ (mg/L)	NO_2^- (mg/L)	NO_3^- (mg/L)	Boron (mg/L)	Waste water load
AX	river K (04.07.2005)	7.96	2.9	6.3	91	0.05	0.04	17.2	0.078	very high
AY	river L (06.11.2004)	8.04	3.1	8.1	134	0.09	0.03	11.2	0.070	medium-high
AZ	river M (27.11.2003)	7.50	2.3	6.2	25	<0.01	<0.01	7.2	n.d.	medium
BA	river N (31.01.2005)	7.84	3.5	9.9	110	0.06	0.12	20.9	0.050	medium
BB	river N (04.07.2005)	7.85	3.3	7.6	133	0.08	0.26	12.6	0.103	medium
BC	river O (31.01.2005)	7.51	0.9	2.5	37	0.25	0.02	5.5	0.010	low
BD	river P (31.01.2005)	8.24	1.4	3.0	235	0.37	0.16	14.1	0.030	low
BE	river Q (31.01.2005)	7.48	1.4	3.5	56	0.26	0.12	15.9	0.040	medium
BF	river Q (04.07.2005)	7.59	2.6	6.6	53	0.04	0.14	13.9	0.033	medium
BG	river R (31.01.2005)	7.69	2.3	6.3	87	0.24	0.10	17.3	0.040	medium-high
BH	river R (04.07.2005)	7.71	4.1	9.6	131	0.19	0.02	15.7	0.079	medium-high
BI	river S (30.01.2005)	7.96	3.2	7.5	198	0.06	0.08	13.3	0.170	high
BJ	river S (04.07.2005)	8.10	4.0	9.3	176	0.02	0.05	7.9	0.147	high
BK	river T (31.01.2005)	7.91	1.5	3.8	119	0.06	0.06	11.5	0.030	medium
BL	river U (31.01.2005)	7.50	1.4	4.1	47	0.13	0.06	7.7	0.020	low
BM	river V (29.06.2005)	8.04	2.8	7.1	258	0.07	0.35	22.9	0.118	high
BN	river W (30.06.2005)	7.37	3.8	13.3	33	0.08	0.11	2.9	0.021	low
BO	river X (30.06.2005)	7.28	1.3	3.9	22	0.04	0.02	2.8	0.010	low
BP	river Y (02.07.2005)	8.39	3.8	10.5	244	0.07	0.21	16.0	0.033	medium
BQ	river Z (02.07.2005)	8.15	7.8	28.0	221	0.05	0.13	15.6	0.034	medium-high
BR	river AA (02.07.2005)	8.33	7.2	27.5	201	0.08	0.16	8.4	0.011	low
BS	river AB (03.07.2005)	7.76	1.8	4.5	56	0.01	0.10	12.4	0.054	medium
BT	river AC (03.07.2005)	8.02	3.6	9.6	128	0.06	0.08	13.7	0.059	low-medium
BU	creek A (07.11.2004)	6.95	2.2	5.9	42	<0.01	<0.01	12.1	<0.01	none

(continued)

Table 4.1 (Continued)

	Water type (sampling date)	pH	TOC (mg/L)	UV_{254nm} (m^{-1})	Alkalinity (mg/L $CaCO_3$)	NH_4^+ (mg/L)	NO_2^- (mg/L)	NO_3^- (mg/L)	Boron (mg/L)	Waste water load
BV	creek A (03.07.2005)	7.45	1.6	3.8	28	0.10	0.04	17.7	0.012	none
BW	creek B (03.07.2005)	7.43	1.4	3.1	18	<0.01	<0.01	14.1	0.017	none
BX	spring A (25.11.2004)	6.73	0.7	0.8	10	<0.01	<0.01	4.5	<0.01	none
BY	spring A (03.07.2005)	7.25	1.1	1.9	19	0.01	0.01	10.0	0.010	none
BZ	spring B (03.07.2005)	6.01	1.1	1.3	14	0.10	0.01	13.7	0.021	none
CA	lake A (25.11.2004)	3.92	49	248	n.d.	0.02	<0.01	1.0	<0.01	none
CB	lake A (03.07.2005)	4.28	21	119	n.d.	0.37	<0.01	<0.50	0.005	none
CC	lake B (03.07.2005)	8.12	1.5	2.9	117	0.08	0.02	3.6	0.021	low-medium
CD	groundwater A (28.09.2004)	n.d.	1.3	2.1	266	<0.01	<0.01	8.3	n.d.	none
CE	groundwater B (27.09.2004)	n.d.	4.5	12.2	437	<0.01	<0.01	<0.5	n.d.	none
CF	groundwater C (11.01.2005)	7.59	0.76	0.9	253	<0.01	<0.01	27.2	0.07	none
CG	groundwater D (11.01.2005)	7.33	0.83	1.1	223	<0.01	<0.01	16.7	0.10	none
CH	groundwater E (26.01.2005)	7.40	0.92	1.1	164	0.01	0.02	3.4	0.09	none
CI	groundwater F (26.01.2005)	7.33	0.78	1.3	236	<0.01	<0.01	23.3	0.09	none
CJ	groundwater G (26.01.2005)	7.46	0.97	1.3	165	0.06	<0.01	<1.0	0.08	none
CK	groundwater H (10.01.2005)	7.29	0.80	1.0	191	<0.01	<0.01	39.8	0.07	none
CL	groundwater I (10.01.2005)	7.16	0.75	0.9	271	<0.01	<0.01	14.4	0.05	none
CM	groundwater J (26.01.2005)	7.14	0.88	1.3	277	<0.01	<0.01	13.8	0.06	none
CN	groundwater K (26.01.2005)	7.40	0.76	1.0	244	<0.01	<0.01	18.3	0.12	none
CO	groundwater L (04.11.2004)	7.40	0.9	1.2	111	<0.01	<0.01	12.0	0.08	influenced
CP	groundwater M (04.11.2004)	7.42	1.2	2.9	153	<0.01	<0.01	11.1	0.08	influenced
CQ	groundwater M (17.01.2005)	7.47	1.3	2.8	163	0.01	<0.01	15.0	0.07	influenced
CR	groundwater M (04.07.2005)	7.87	1.0	1.5	163	<0.01	<0.01	9.9	0.067	influenced
CS	groundwater M (11.07.2005)	7.45	1.1	1.5	163	<0.01	<0.01	10.2	0.070	influenced

(n.d.: not determined; concentrations of ammonia, nitrite and nitrate given as mg/L NH_4^+, mg/L NO_2^-, and mg/L NO_3^-)

Table 4.2
Nitrosamine formation potential of investigated waters

	Water type (sampling date)	*NDMA (ng/L)*	*NEMA (ng/L)*	*NDEA (ng/L)*	*NDPA (ng/L)*	*NDBA (ng/L)*	*NPIP (ng/L)*	*NPYR (ng/L)*	*NMOR (ng/L)*	*NdcHxA (ng/L)*
A	**river A, I** *(31.01.2005)*	**34**	<1	<1	<1	<1	<1	**1.1**	<1	<1
B	**river A, II** *(10.09.2004)*	**24**	<1	<1	<1	<1	<1	**1.1**	<1	<1
C	**river A, II** *(08.10.2004)*	**23**	<1	<1	<1	<1	<1	**1.3**	<1	<1
D	**river A, II** *(22.10.2004)*	**24**	<1	**4.2**	<1	<1	<1	**2.0**	<1	<1
E	**river A, II** *(01.11.2004)*	**32**	<1	**3.3**	<1	<1	<1	<1	<1	<1
F	**river A, II** *(31.01.2005)*	**90**	<1	<1	<1	<1	<1	**1.5**	<1	<1
G	**river A, II** *(07.06.2005)*	**1000***	<1	**1.5**	<1	<1	<1	**1.5**	<1	<1
H	**river A, II** *(29.06.2005)*	**735***	<1	**1.2**	<1	<1	<1	**2.0**	<1	<1
I	**river A, III** *(30.10.2004)*	**61**	<1	**1.4**	<1	<1	<1	**1.2**	<1	<1
J	**river A, IV** *(30.10.2004)*	**72**	<1	**2.5**	<1	<1	<1	**2.6**	**2.1**	<1
K	**river A, V** *(30.10.2004)*	**77**	<1	**5.8**	<1	<1	<1	**2.6**	<1	<1
L	**river A, VI** *(04.07.2005)*	**440***	<1	<1	<1	<1	<1	**2.3**	**1.3**	<1
M	**river A, VII** *(08.11.2004)*	**37**	<1	<1	<1	<1	<1	**1.5**	<1	<1
N	**river A, VII** *(04.07.2005)*	**410***	<1	<1	<1	<1	<1	**3.9**	**1.0**	<1
O	**river A, VIII** *(22.10.2004)*	**50**	<1	**1.2**	<1	<1	<1	**1.9**	<1	<1
P	**river A, IX** *(04.07.2005)*	**450***	<1	**2.5**	<1	<1	<1	**4.7**	<1	<1
Q	**river A, X** *(09.09.2004)*	**47**	<1	<1	<1	<1	<1	**1.9**	<1	<1
R	**river A, X** *(04.11.2004)*	**54**	<1	**2.4**	<1	<1	<1	**2.0**	<1	<1
S	**river A, X** *(17.01.2005)*	**85**	<1	**5.5**	<1	<1	<1	**2.2**	<1	<1
T	**river A, X** *(30.01.2005)*	**51**	<1	**2.0**	<1	<1	<1	**2.1**	<1	<1
U	**river A, X** *(04.07.2005)*	**390***	<1	**1.7**	<1	<1	<1	**3.4**	**1.0**	<1
V	**river A, X** *(11.07.2005)*	**490***	<1	**2.5**	<1	<1	<1	**2.8**	**1.0**	<1
W	**river A, XI** *(04.11.2004)*	**72**	<1	**1.8**	<1	<1	<1	**2.1**	<1	<1
X	**river A, XI** *(04.07.2005)*	**409***	<1	**1.7**	<1	<1	<1	**4.6**	**1.4**	<1

(continued)

Table 4.2 (Continued)

	Water type (sampling date)	*NDMA (ng/L)*	*NEMA (ng/L)*	*NDEA (ng/L)*	*NDPA (ng/L)*	*NDBA (ng/L)*	*NPIP (ng/L)*	*NPYR (ng/L)*	*NMOR (ng/L)*	*NdcHxA (ng/L)*
Y	***river A, XII*** *(06.11.2004)*	**60**	<1	**1.5**	<1	<1	<1	**2.6**	<1	<1
Z	***river A, XII*** *(04.07.2005)*	**360***	<1	**1.3**	<1	<1	<1	**5.7**	**1.2**	<1
AA	***river A, XIII*** *(06.11.2004)*	**56**	<1	**1.3**	<1	<1	<1	**3.1**	<1	<1
AB	***river A, XIV*** *(06.11.2004)*	**56**	<1	**2.0**	<1	<1	<1	**2.5**	<1	<1
AC	***river B, I*** *(26.10.2004)*	**66**	<1	**2.5**	<1	<1	<1	**3.0**	<1	<1
AD	***river B, II*** *(26.10.2004)*	**87**	<1	**1.2**	<1	<1	**1.9**	**5.6**	**1.6**	<1
AE	***river B,III*** *(04.07.2005)*	**182**	<1	**1.0**	<1	<1	<1	**5.1**	**1.1**	<1
AF	***river C*** *(26.10.2004)*	**48**	<1	<1	<1	<1	<1	**2.5**	<1	<1
AG	***river D*** *(06.11.2004)*	**362**	**1.4**	<1	**7.5**	<1	**5.9**	**35**	**4.0**	<1
AH	***river D*** *(30.01.2005)*	**357**	**2.2**	**1.5**	<1	**2.5**	**28**	**34**	**6.0**	<1
AI	***river D*** *(04.07.2005)*	**326**	**1.8**	**1.2**	<1	<1	**2.5**	**27**	**6.8**	<1
AJ	***river E*** *(06.11.2004)*	**96**	<1	<1	<1	<1	<1	**4.9**	<1	<1
AK	***river E*** *(30.01.2005)*	**66**	<1	<1	<1	<1	<1	**3.6**	**2.1**	<1
AL	***river E*** *(04.07.2005)*	**90**	<1	<1	<1	<1	**1.1**	**6.8**	**1.8**	<1
AM	***river F, I*** *(03.07.2005)*	**64**	<1	<1	<1	<1	<1	**2.1**	<1	<1
AN	***river F , II*** *(04.11.2004)*	**61**	<1	<1	<1	<1	<1	**3.7**	<1	<1
AO	***river F , III*** *(30.01.2005)*	**40**	<1	<1	<1	<1	<1	**1.7**	<1	<1
AP	***river G*** *(04.11.2004)*	**58**	<1	**1.3**	<1	<1	**1.4**	**6.0**	**1.5**	<1
AQ	***river G*** *(30.01.2005)*	**34**	<1	<1	<1	<1	<1	**2.5**	<1	<1
AR	***river H*** *(06.09.2004)*	**57**	<1	<1	<1	<1	<1	**5.0**	<1	<1
AS	***river H*** *(29.11.2004)*	**58**	<1	<1	<1	<1	<1	<1	<1	<1
AT	***river H*** *(04.07.2005)*	**71**	<1	**1.0**	<1	<1	<1	**5.7**	**1.0**	<1
AU	***river I*** *(25.11.2004)*	**14**	<1	<1	<1	<1	<1	<1	<1	<1
AV	***river J*** *(25.11.2004)*	**21**	<1	<1	<1	<1	<1	<1	<1	<1
AW	***river K*** *(08.11.2004)*	**115**	<1	**1.5**	<1	<1	**1.5**	**4.8**	<1	<1

(continued)

Table 4.2 (Continued)

	Water type (sampling date)	*NDMA (ng/L)*	*NEMA (ng/L)*	*NDEA (ng/L)*	*NDPA (ng/L)*	*NDBA (ng/L)*	*NPIP (ng/L)*	*NPYR (ng/L)*	*NMOR (ng/L)*	*NdcHxA (ng/L)*
AX	***river K*** *(04.07.2005)*	**120**	<1	<1	<1	<1	<1	**6.6**	**1.0**	<1
AY	***river L*** *(06.11.2004)*	**89**	<1	**1.7**	**1.1**	<1	<1	**3.0**	<1	<1
AZ	***river M*** *(27.11.2003)*	**21**	<1	<1	<1	<1	<1	**1.6**	<1	<1
BA	***river N*** *(31.01.2005)*	**20**	<1	<1	<1	<1	<1	<1	<1	<1
BB	***river N*** *(04.07.2005)*	**38**	<1	<1	<1	<1	<1	**2.5**	<1	<1
BC	***river O*** *(31.01.2005)*	**12**	<1	<1	<1	<1	<1	<1	<1	<1
BD	***river P*** *(31.01.2005)*	**15**	<1	<1	<1	<1	<1	**1.4**	<1	<1
BE	***river Q*** *(31.01.2005)*	**42**	<1	**23**	<1	<1	<1	**2.0**	<1	<1
BF	***river Q*** *(04.07.2005)*	**36**	<1	<1	<1	<1	<1	**1.4**	<1	<1
BG	***river R*** *(31.01.2005)*	**54**	<1	<1	<1	<1	<1	**2.2**	<1	<1
BH	***river R*** *(04.07.2005)*	**95**	<1	<1	<1	<1	<1	**6.5**	<1	<1
BI	***river S*** *(30.01.2005)*	**61**	<1	<1	<1	<1	<1	**3.4**	<1	<1
BJ	***river S*** *(04.07.2005)*	**53**	<1	<1	<1	<1	<1	**4.9**	<1	<1
BK	***river T*** *(31.01.2005)*	**19**	<1	<1	<1	<1	<1	**1.0**	<1	<1
BL	***river U*** *(31.01.2005)*	**19**	<1	<1	<1	<1	<1	<1	<1	<1
BM	***river V*** *(29.06.2005)*	**107**	<1	<1	<1	<1	<1	**7.9**	<1	<1
BN	***river W*** *(30.06.2005)*	**52**	<1	<1	<1	<1	<1	**2.7**	<1	<1
BO	***river X*** *(30.06.2005)*	**8.2**	<1	<1	<1	<1	<1	< 1	<1	<1
BP	***river Y*** *(02.07.2005)*	**38**	<1	<1	<1	<1	<1	**2.8**	<1	<1
BQ	***river Z*** *(02.07.2005)*	**74**	<1	<1	<1	<1	**1.3**	**5.5**	<1	<1
BR	***river AA*** *(02.07.2005)*	**37**	<1	<1	<1	<1	**1.2**	**1.0**	<1	<1
BS	***river AB*** *(03.07.2005)*	**40**	<1	<1	<1	<1	<1	**1.8**	<1	<1
BT	***river AC*** *(03.07.2005)*	**48**	<1	<1	<1	<1	<1	**1.0**	<1	<1
BU	***creek A*** *(07.11.2004)*	**5.1**	<1	<1	<1	<1	<1	<1	<1	<1

(continued)

Table 4.2 (Continued)

	Water type (sampling date)	*NDMA (ng/L)*	*NEMA (ng/L)*	*NDEA (ng/L)*	*NDPA (ng/L)*	*NDBA (ng/L)*	*NPIP (ng/L)*	*NPYR (ng/L)*	*NMOR (ng/L)*	*NdcHxA (ng/L)*
BV	***creek A*** *(03.07.2005)*	**7.1**	<1	<1	<1	<1	<1	**2.1**	<1	<1
BW	***creek B*** *(03.07.2005)*	**4.4**	<1	<1	<1	<1	<1	< 1	<1	<1
BX	***spring A*** *(25.11.2004)*	**2.4**	<1	<1	<1	<1	<1	<1	<1	<1
BY	***spring A*** *(03.07.2005)*	**2.9**	<1	<1	<1	<1	<1	**1.7**	<1	<1
BZ	***spring B*** *(03.07.2005)*	**6.9**	<1	<1	<1	<1	<1	**2.5**	<1	<1
CA	***lake A*** *(25.11.2004)*	**1.7**	<1	<1	<1	<1	<1	<1	<1	<1
CB	***lake A*** *(03.07.2005)*	**5.6**	<1	<1	<1	<1	<1	**1.4**	<1	<1
CC	***lake B*** *(03.07.2005)*	**17**	<1	<1	<1	<1	<1	**2.1**	<1	<1
CD	***groundwater A*** *(28.09.2004)*	**5.9**	<1	<1	<1	<1	<1	<1	<1	<1
CE	***groundwater B*** *(27.09.2004)*	**8.5**	<1	<1	<1	<1	<1	<1	<1	<1
CF	***groundwater C*** *(11.01.2005)*	**1.0**	<1	<1	<1	<1	<1	<1	<1	<1
CG	***groundwater D*** *(11.01.2005)*	**1.5**	<1	<1	<1	<1	<1	<1	<1	<1
CH	***groundwater E*** *(26.01.2005)*	**1.7**	<1	<1	<1	<1	<1	<1	<1	<1
CI	***groundwater F*** *(26.01.2005)*	**1.4**	<1	<1	<1	<1	<1	<1	<1	<1
CJ	***groundwater G*** *(26.01.2005)*	**1.6**	<1	<1	<1	<1	<1	<1	<1	<1
CK	***groundwater H*** *(10.01.2005)*	**1.2**	<1	<1	<1	<1	<1	<1	<1	<1
CL	***groundwater I*** *(10.01.2005)*	**2.1**	<1	<1	<1	<1	<1	<1	<1	<1
CM	***groundwater J*** *(26.01.2005)*	**2.5**	<1	<1	<1	<1	<1	<1	<1	<1
CN	***groundwater K*** *(26.01.2005)*	**2.3**	<1	<1	<1	<1	<1	<1	<1	<1
CO	***groundwater L*** *(04.11.2004)*	**5.3**	<1	<1	<1	<1	<1	<1	<1	<1
CP	***groundwater M*** *(04.11.2004)*	**16**	<1	**1.1**	<1	<1	<1	<1	<1	<1
CQ	***groundwater M*** *(17.01.2005)*	**12**	<1	<1	<1	<1	<1	<1	<1	<1
CR	***groundwater M*** *(04.07.2005)*	**27**	<1	<1	<1	<1	<1	<1	<1	<1
CS	***groundwater M*** *(11.07.2005)*	**23**	<1	<1	<1	<1	<1	<1	<1	<1

Conditions: 0.4 mM chloramine, pH 7, 20°C, 168 h

* data excluded from evaluation concerning boron correlation

It can be seen from the data in Table 4.2 that NDMA was the main nitrosamine formed during chloramination in all waters investigated. NPYR and NDEA were the nitrosamines second and third most often found, however, levels were always at least by an order of magnitude lower than those of NDMA. In surface waters showing an extremely high NDMA formation potential, low levels of NEMA, NDPA and NPIP were also detected occasionally. Typically, these surface waters were also characterized by a high (estimated) wastewater load. On the other hand, surface waters without any waste water portion (e.g. samples BU, BV, BW, BX, BY, BZ, CA, and CB) showed only little potential for NDMA formation during chloramination. A concise presentation of the results for the surface waters (i.e. river and lake water samples) based on a statistical evaluation of the data in Table 4.2 is given in Figure 4.6.

Investigated groundwaters all showed rather low NDMA levels following chloramination when compared to surface waters. Within groundwater samples, the waters being influenced by surface water (samples CO, CP, CQ, CR, CS) and those with a higher TOC level (samples CD and CE) seem to be more susceptible to NDMA formation during chloramination than non-influenced waters with lower TOC levels. Furthermore, one of the investigated surface water samples without waste water portion was characterized by an outstanding TOC level of 49 mg/L due to a high content of natural organic matter (NOM). This high natural NOM content, however, leads to a NDMA formation of only 1.7 ng/L.

A comprehensive evaluation of the present data set leads to the hypothesis that the precursor portion within the NOM is rather low and most probably only of secondary importance for NDMA formation during chloramination. To support this hypothesis, boron levels were determined in most samples as an indicator for pollution of anthropogenic origin. Besides the generally low natural boron content of surface waters, boric compounds are applied frequently in industries (e.g. enamel and glass industries, galvanics, photo chemistry, production of mineral dyestuffs and impregnations), but are predominantly used as detergent additives (e.g. sodium perborate as bleaching agent) and thus are mainly released via municipal wastewater treatment plants in the aquatic environment. Figure 4.7 shows the interesting correlation between NDMA formation and boron content found for surface waters.

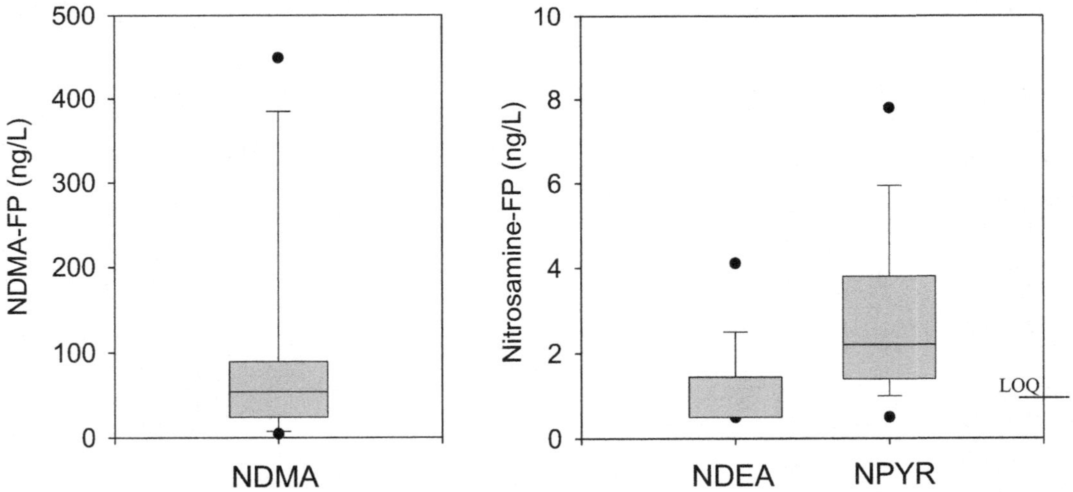

Conditions: 0.4 mM chloramine, pH 7, 20 °C, 168 h (n = 81 samples, 34 different rivers and lakes)

Note: Box indicates 25 and 75%ile, whiskers 10 and 90%ile and dots 5 and 95%ile

Figure 4.6 Nitrosamine formation potential of German surface waters

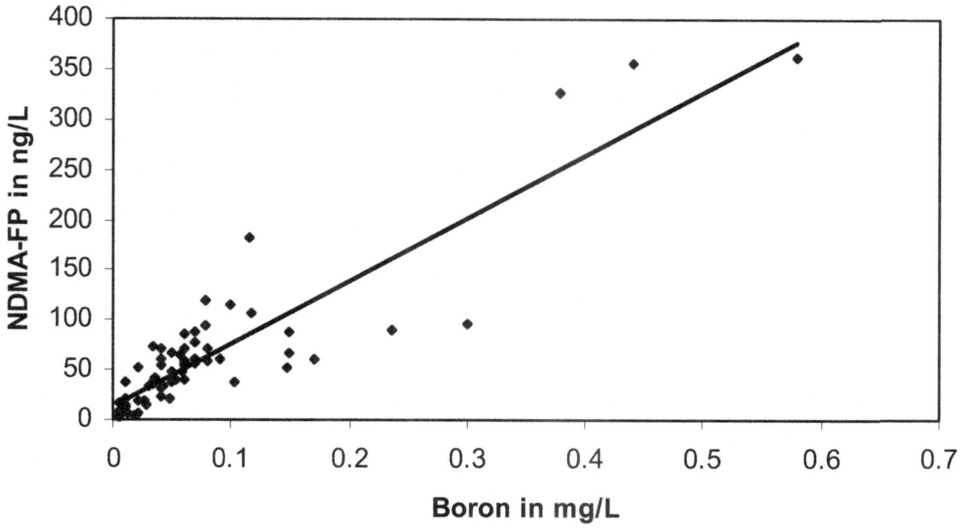

Figure 4.7 Correlation between boron concentration and NDMA formation in a NFP test

The rather good correlation between boron concentration and NDMA formation potential (correlation coefficient r = 0.90) underlines the importance of anthropogenic water constituents for NDMA formation during chloramination. None of the other parameters analyzed in the basic set of physical-chemical parameters showed a comparable good correlation to the NFP. Data of samples taken in June and July 2005 from river A were excluded from this evaluation, since the extraordinary high levels of NDMA precursors found at these dates were caused by an abnormal occurrence at an industrial wastewater treatment plant in the upper part of the river due to a full load situation with release of certain amounts of untreated wastewater.

NFP tests were also performed with waters sampled within ten days in the course of one river, i.e. with increasing wastewater load (river A). In principle, Figure 4.8 shows the increasing trend of the NDMA and NPYR formation potential along the river. Certain sampling points along the river, however, demonstrated a higher potential of NDMA formation than expected from the general trend. These sampling points (III, IV, V, and XI) are typically influenced by tributaries or discharges of major (industrial) waste water treatment plants and thus show deviating characteristics due to the incomplete mixing with the main water body.

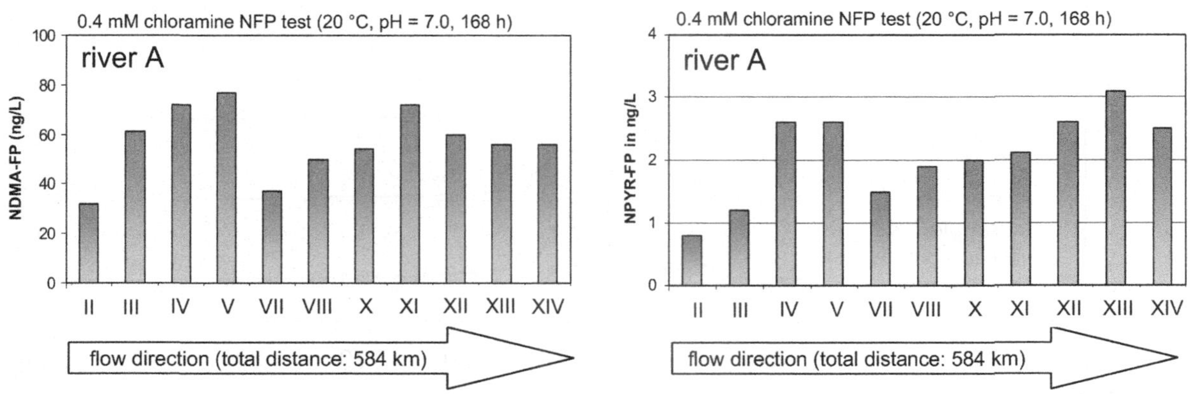

Figure 4.8 Formation potential for NDMA and NPYR in the course of a river with increasing wastewater load (sampling time October/November 2004)

For further characterization of the organic matrix in six selected water samples (3 samples with a high and 3 samples with a low NDMA formation potential), the LC-OCD-OND technique was used (Huber and Frimmel 1992a, 1992b). LC stands for liquid chromatography, OCD for organic carbon detection, and OND for organic nitrogen detection. All experiments were performed in collaboration with Dr. Stefan Huber (DOC-Labor Dr. Huber, Karlsruhe, Germany). The respective results are given in Figure 4.9 and Table 4.3. An explanation of the relevant parameters is given below Table 4.3.

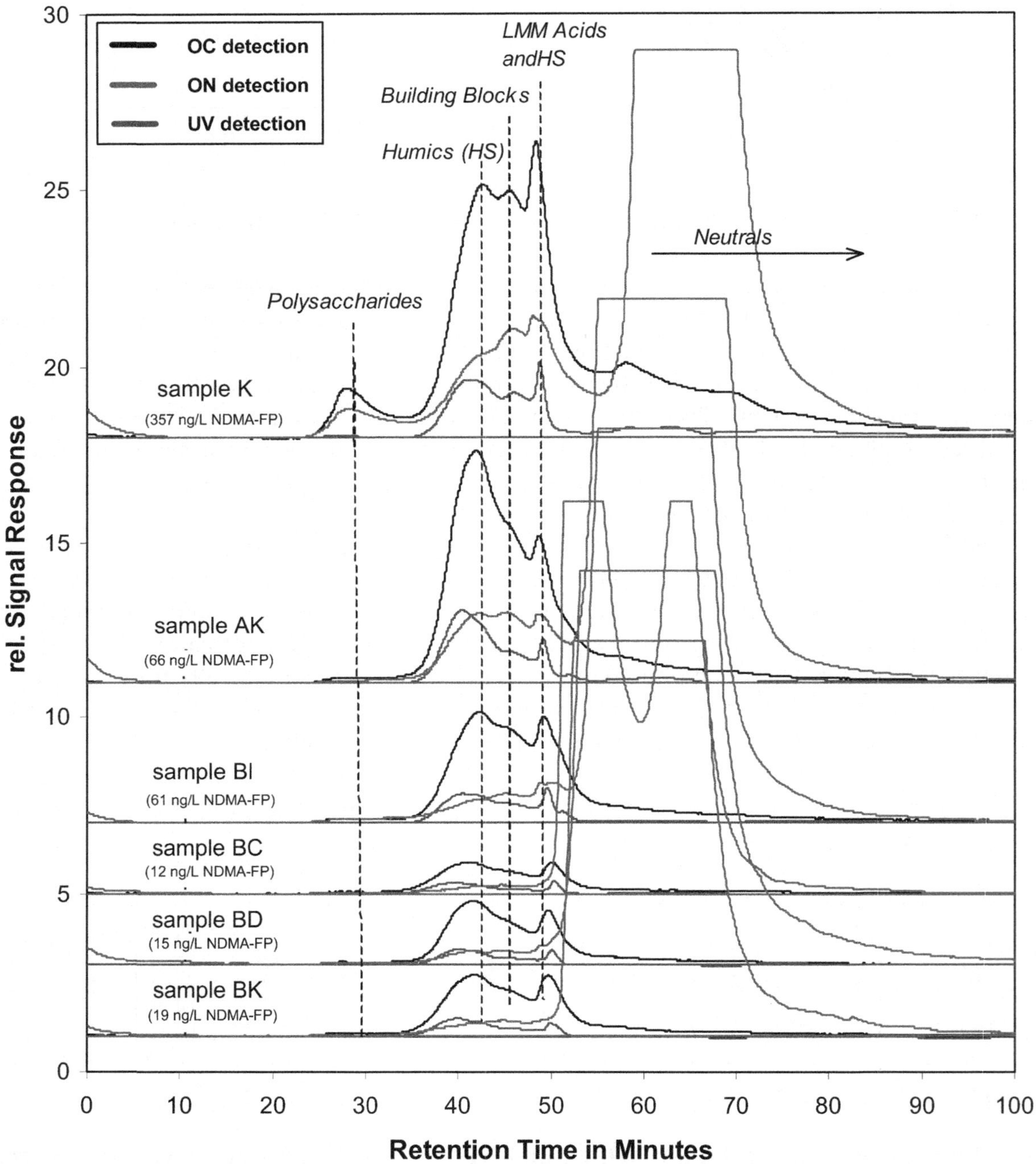

Figure 4.9 LC-OCD-OND chromatograms of selected surface water samples

Table 4.3

Results of LC-OCD-OND measurements of selected surface water samples

	Partitioning of Organic Carbon (OC)					Chromatographic Fractionation of Organic Carbon (CDOC)									(UV@254 nm)	
	Approx. Molecular Weights in g/mol: →					>>20.000		~1000 (see separate HS-Diagram)				300-500	<350	<350		
	TOC=DOC+POC DOC=CDOC+HOC; Note: POC, hence TOC may be too low															
	TOC	DOC	POC	HOC	CDOC	Poly-saccha-ridic	DON	Humic Subst. (HS)	DON	Aromaticity (SUVA-HS)	Mol-Weight (Mn)	Building Blocks	Neutrals	Acids	Inorg. Colloid.	SUVA
	total OC	dissolved	particul.	hydrophob.	hydrophil.		(Norg)		(Norg)						SAC	(SAC/OC)
	ppb-C	ppb-C	ppb-C	ppb-C	ppb-C	ppb-C	ppb-N	ppb-C	ppb-N	L/(mg*m)	g/mol	ppb-C	ppb-C	ppb-C	(m^{-1})	L/(mg*m)
	% TOC	% TOC	% TOC	% TOC	% TOC	% TOC	--	% TOC	--	--	--	% TOC	% TOC	% TOC	--	--
sample K (357 ng/L NDMA-FP)	**8298**	**8254**	**44**	**1743**	**6511**	**434**	41	**2612**	133	**2,59**	**684**	**1317**	**2146**	**3**	**0,15**	**2,19**
	100	99,5	0,5	21,0	78,5	5,2	--	31,5	--	--	--	15,9	25,9	0,0	--	--
sample AK (66 ng/L NDMA-FP)	**4474**	**4336**	**137**	**575**	**3761**	**74**	5	**2204**	108	**3,33**	**743**	**784**	**700**	**0**	**0,05**	**2,73**
	100	96,9	3,1	12,9	84,1	1,6	--	49,3	--	--	--	17,5	15,6	0,0	--	--
sample BI (61 ng/L NDMA-FP)	**2823**	**2707**	**116**	**416**	**2292**	**46**	7	**1314**	40	**3,18**	**665**	**458**	**474**	**0**	**0,05**	**2,74**
	100	95,9	4,1	14,7	81,2	1,6	--	46,6	--	--	--	16,2	16,8	0,0	--	--
sample BC (12 ng/L NDMA-FP)	**906**	**808**	**97**	**106**	**703**	**23**	0	**436**	12	**3,25**	**711**	**97**	**147**	**0**	**0,09**	**3,03**
	100	89,3	10,7	11,7	77,6	2,5	--	48,1	--	--	--	10,7	16,3	0,0	--	--
sample BD (15 ng/L NDMA-FP)	**1420**	**1274**	**146**	**132**	**1142**	**35**	3	**725**	22	**2,32**	**718**	**192**	**191**	**0**	**0,03**	**2,38**
	100	89,7	10,3	9,3	80,4	2,4	--	51,0	--	--	--	13,5	13,4	0,0	--	--
sample BK (19 ng/L NDMA-FP)	**1498**	**1444**	**54**	**190**	**1254**	**32**	1	**736**	24	**2,97**	**723**	**223**	**234**	**29**	**0,02**	**2,64**
	100	96,4	3,6	12,7	83,7	2,1	--	49,2	--	--	--	14,9	15,6	1,9	--	--

SUMMARIC PARAMETERS:

TOC (Total OC): Determined in the column bypass. Underestimation of TOC-values is possible for samples rich in particulate bound OC (POC) due to incomplete oxidation.
DOC (Dissolved OC): Like for TOC, but a small 0.45 µm filter is inserted automatically in the sample stream.
POC (Particulate OC): Calculated as difference TOC – DOC.
HOC (Hydrophobic OC): Calculated as difference TOC – CDOC (CDOC = Chromatographic DOC). CDOC is the OC value obtained by area integration of the total chromatogram. Therefore, all OC retained on the column is defined as "hydrophobic".
SUVA (SAC/DOC): Additional parameter derived from DOC and SAC.

NATURAL ORGANIC MATTER (NOM):

Humics (HS): In LC-OCD measurements there is a tight definition for HS based on retention time, peak shape and SAC. Calibration was performed on the basis of "Suwannee River" Standards IHSS-FA and IHSS-HA. In addition, statistical data are given, like molecular mass distribution (Mn) and aromaticity (SAC/OC).
Building Blocks (HS-Hydrolysates): The HS fraction is overlain by broad shoulders. Shape, concentration and UV activity varies. The shoulders can be produced from HS by ultrasonification or mild oxidation. This suggests that the shoulders are sub-units ("building blocks") of HS, molecular weights between 300-450 g/mol. Building Blocks are perhaps weathering and oxidation products of HS, they cannot be removed in flocculation processes.
LMW Organic-Acids: In this fraction all aliphatic low-molecular mass organic acids co-elute due to an ion-chromatographic effect. A small amount of HS may fall into this fraction and has to be subtracted on the basis of SAC/OC ratios.
LMW Neutrals: According to theory, only low-molecular weight weakly charged hydrophilic or slightly hydrophobic („amphiphilic") compounds appear in this fraction, like alcohols, aldehydes, ketones, amino acids. The hydrophobic character increases with retention time, e.g. pentanol at 120 min, octanol at 240 min. Compounds not eluting before 200 min are rated „hydrophobic" (HOC)
Polysaccharides („EPS", including amino sugars, polypeptides and proteins): This fraction is very high in molecular weight (100.000-2 Mio. g/mol), hydrophilic, not UV-absorbing.
Inorganic Colloids (found only in UV chromatograms): Negatively charged inorganic polyelectrolytes, polyhydroxides and oxihydrates of Fe, Al, Si are present here and are detected by UV light-scattering (Raleigh-effect).

Using the LC-OCD-OND technique, samples are analyzed without any further pre-treatment. First, natural organic matter (NOM) is fractionated by size exclusion chromatography (SEC) and then the organic carbon and organic nitrogen content of the separated fractions is detected as carbon dioxide and as nitrate, respectively, after photo-oxidation in a thin-film reactor. Fractionation of NOM during SEC is based on three effects, size exclusion, ionic and hydrophobic interactions, making the interpretation of the chromatograms rather complex. Interpretations given in this report are mainly based on long-lasting experiences by Stefan Huber.

Samples analyzed by the LC-OCD-OND method were samples K, AK, BI (high NDMA-formation potential) and samples BC, BD, BK (low NDMA-formation potential). According to the results, the NDMA formation potential might correlate with several parameters. Major similarities were found for hydrophobic organic carbon (HOC), the DON in the polysaccharide fraction, and the organic carbon in the low molecular weight (LMW) neutrals fraction. In particular, the result for the polysaccharidic DON fraction is interesting. The major difference found in the NDMA formation potential for sample K (357 ng/L NDMA) and sample AK (66 ng/L NDMA) corresponds quite well to the difference in the polysaccharidic DON concentration with 41 ppb N and 5 ppb N, respectively.

Several samples investigated by the LC-OCD-OND technique were analyzed in parallel for aliphatic amines. The analytical amine method (see Chapter 2) allows for determination of dimethylamine (DMA), pyrrolidine (PYR) and other amines at levels of a few ng/L. Typically reporting limits are at 15 or 30 ng/L. However, in the case of dimethylamine and pyrrolidine findings below this level are also reported for research purposes in brackets, since they might help in comparing the nitrosamine formation potential with the corresponding amine levels (see Table 4.4).

Figure 4.10 illustrates the concentrations of secondary amines relevant for nitrosamine formation in a condensed form. As can be seen from the box plot presentation, concentrations of these amines rarely exceed levels of 200 ng/L per single compound. Based on the conversion rates determined in laboratory-based experiments, concentrations of 200 ng/L amine would be capable to form about 8.5 ng/L NDMA, 1.4 ng/L NDEA, 5.1 ng/L NPYR, or 0.3 ng/L NMOR at a conversion rate of 2.6%, 0.5%, 1.8%, and 0.1%, respectively. From these calculations it is evident that at least NDMA and NPYR formation during chloramination is mainly based on other amine precursors. Levels of formed NMOR and NDEA, however, are in a range that might be caused by the levels of the secondary amine in the chloraminated water. Pairwise comparison of amine levels and the corresponding nitrosamine levels formed during chloramination in each sample revealed apparent conversion rates that were on an average about 50fold higher for DMA than the experimentally determined one, 5fold higher for DEA, 9fold higher for PYR, and 6fold higher for MOR. This indicates that also for NDEA, NPYR, and NMOR formation other precursor compounds amines might be of relevance.

Since conversion rates of DMA into NDMA by chloramination can not be generalized for all amines with dimethylamine function, a set of tertiary amines was selected for chloramination experiments in order to study the influence of various structural elements of tertiary amines located next to the dimethylamine function. The molecular structures of these compounds are shown in Figures 4.11 and 4.12 and the determined conversion rates are summarized in Table 4.5.

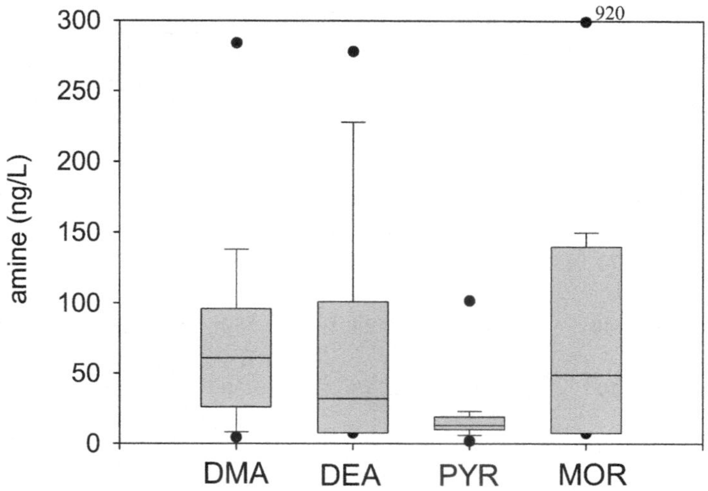

Note: Box indicates 25 and 75%ile, whiskers 10 and 90%ile and dots 5 and 95%ile

Figure 4.10 Amine concentrations in German surface waters (31 samples from 25 rivers and lakes)

Table 4.4
Levels of aliphatic amines in the waters under investigation

	Water type (*sampling date*)	*MMA (ng/L)*	*DMA (ng/L)*	*MEA (ng/L)*	*DEA (ng/L)*	*DBA (ng/L)*	*PIP (ng/L)*	*PYR (ng/L)*	*MOR (ng/L)*
F	*river A, II* (*31.01.2005*)	n.d.	n.d.	n.d.	n.d.	n.d.	n.d.	n.d.	n.d.
G	*river A, II* (*07.06.2005*)	n.d.	n.d.	n.d.	n.d.	n.d.	n.d.	n.d.	n.d.
H	*river A, II* (*29.06.2005*)	**32**	**130**	**100**	**290**	<30	<15	(9)	**52**
L	*river A, VI* (*04.07.2005*)	**54**	**78**	**39**	**79**	<30	<15	(10)	**120**
N	*river A, VII* (*04.07.2005*)	**94**	**96**	**75**	**120**	<30	**24**	**23**	**150**
P	*river A, IX* (*04.07.2005*)	**45**	**83**	**31**	**250**	<30	<15	**23**	**130**
T	*river A, X* (*30.01.2005*)	n.d.	n.d.	n.d.	n.d.	n.d.	n.d.	n.d.	n.d.
U	*river A, X* (*04.07.2005*)	**59**	**120**	<30	**110**	<30	<15	<15	**140**
V	*river A, X* (*11.07.2005*)	n.d.	n.d.	n.d.	n.d.	n.d.	n.d.	n.d.	n.d.
X	*river A, XI* (*04.07.2005*)	**63**	**90**	<30	**140**	<30	<15	**22**	**140**
Z	*river A, XII* (*04.07.2005*)	**49**	**79**	<30	**70**	<30	<15	**23**	**145**
AE	*river B,III* (*04.07.2005*)	**67**	**58**	**36**	**270**	<30	<15	(12)	**140**
AI	*river D* (*04.07.2005*)	**420**	**410**	<30	**140**	**53**	**71**	**217**	**1000**

(continued)

Table 4.4 (continued)

	Water type (*sampling date*)	*MMA (ng/L)*	*DMA (ng/L)*	*MEA (ng/L)*	*DEA (ng/L)*	*DBA (ng/L)*	*PIP (ng/L)*	*PYR (ng/L)*	*MOR (ng/L)*
AL	***river E*** (*04.07.2005*)	**120**	**120**	**39**	**69**	**46**	**15**	**18**	**850**
AM	***river F, I*** (*03.07.2005*)	**65**	**57**	<30	**46**	<30	<15	(13)	**49**
AT	***river H*** (*04.07.2005*)	**81**	**72**	**151**	**101**	**58**	<15	(12)	**150**
AX	***river K*** (*04.07.2005*)	**110**	**140**	**39**	**24**	<30	**26**	**18**	**130**
BB	***river N*** (*04.07.2005*)	**57**	**29**	<30	**15**	<30	<15	(10)	**87**
BF	***river Q*** (*04.07.2005*)	**72**	**30**	<30	<15	<30	<15	(9)	<15
BH	***river R*** (*04.07.2005*)	**220**	**85**	**70**	**32**	<30	<15	**18**	**67**
BJ	***river S*** (*04.07.2005*)	**67**	**35**	<30	<15	<30	<15	**17**	**20**
BM	***river V*** (*29.06.2005*)	**76**	**110**	<30	**57**	**120**	<15	**17**	**45**
BN	***river W*** (*30.06.2005*)	**60**	**28**	<30	**45**	<30	<15	(13)	**100**
BO	***river X*** (*30.06.2005*)	<30	(13)	<30	<15	<30	<15	(6)	<15
BP	***river Y*** (*02.07.2005*)	**56**	**26**	**91**	<15	<30	<15	(10)	<15
BQ	***river Z*** (*02.07.2005*)	**99**	**61**	<30	**26**	<30	<15	**21**	**18**
BR	***river AA*** (*02.07.2005*)	**78**	**32**	<30	<15	<30	<15	(3)	<15
BS	***river AB*** (*03.07.2005*)	**47**	**18**	<30	<15	<30	<15	(13)	<15
BT	***river AC*** (*03.07.2005*)	**54**	(13)	<30	<15	<30	<15	(6)	<15
BV	***creek A*** (*03.07.2005*)	**140**	(6)	**190**	**30**	<30	<15	**16**	<15
BW	***creek B*** (*03.07.2005*)	<30	(13)	<30	<15	<30	<15	(6)	<15
BY	***spring A*** (*03.07.2005*)	<30	(7)	**230**	<15	<30	<15	(13)	<15
BZ	***spring B*** (*03.07.2005*)	**1100**	**70**	**440**	**70**	<30	<15	**25**	<15
CB	***lake A*** (*03.07.2005*)	**240**	**200**	**40**	<15	<30	<15	**19**	<15
CC	***lake B*** (*03.07.2005*)	**31**	(2)	**76**	<15	<30	<15	**15**	<15
CR	***groundwater M*** (*04.07.2005*)	**110**	**72**	<30	**19**	<30	<15	<15	20
CS	***groundwater M*** (*11.07.2005*)	n.d.	n.d.	n.d.	n.d.	n.d.	n.d.	n.d.	n.d.

n.d.: not determined

Table 4.5
Conversion rates of DMA and selected tertiary amines during chloramination

Compound investigated with the NFP test	Level spiked (nM)	NDMA formed (ng/L)	Molar conversion rate (%)
Dimethylamine → NDMA	25	14	0.76
Trimethylamine → NDMA	25	21	1.1
Dimethylformamide → NDMA	25	7.5	0.40
4-(Dimethylamine)azobenzene → NDMA	25	170	9.4
N,N-Dimethyl-4-nitrosoaniline → NDMA	25	14	0.76
N,N,N‘,N‘-Tetramethyl-p-phenylenediamine→ NDMA	12.5	29	1.6
N,N-Dimethyl-p-phenylenediamine → NDMA	25	42	2.3
Methylene blue → NDMA	25	19	1.0
Toluylene blue → NDMA	25	21	1.1
Diuron → NDMA	25	28	1.5
Diphenamid → NDMA	25	9.4	0.51
Oxamyl → NDMA	25	2.5	0.13
Loperamid → NDMA	25	15	0.81
Dimethyldithiocarbamate → NDMA	25	46	2.5
Tetracycline → NDMA	25	23	1.2
Ranitidine → NDMA	25	1200	62.9
5-(Dimethylaminomethyl)furfuryl alcohol → NDMA	25	930	50.3
N,N-Dimethylthiophene-2-methylamine → NDMA	25	920	49.5
Nizatidine → NDMA	25	91	4.9
Gramine → NDMA	25	53	2.9
Daminozide → NDMA	25	2.6	0.14
Dacarbazine → NDMA	25	29	1.6
Cyazofamid → NDMA	25	2.0	0.11
Thiothixene → NDMA	25	1.9	0.10
N,N-Dimethyl-N‘-(4-methylphenyl)-sulfamide → NDMA	25	6.6	0.36

Conditions: 0.4 mM chloramine, pH 7, 20 °C, 168 h (matrix: drinking water)

The data in Table 4.5 clearly demonstrate that nitrosamine formation is substantially enhanced for some tertiary amine compounds providing certain structural characteristics. Among the substances tested, ranitidine, a pharmaceutical often used against gastritis, showed with 62.9% the highest conversion efficiency. Thus, with the same number of molecules ranitidine forms about 83fold more molecules NDMA than does DMA providing a conversion efficiency of 0.76% only. Several tertiary amines show conversion efficiencies comparable to those of dimethylamine and trimethylamine in the range of 0.7 to 3% (e.g. diuron, tetracyclin, methylene blue). However, some tertiary amines even show less conversion than dimethylamine (e.g. oxamyl, cyazofamid, thiothixene). A closer look at the structural characteristics of these compounds reveals in particular two important facts. NDMA formation from a structure with the dimethylamine group next to a carbonyl or a sulfonyl group is typically rather low, while structures with the dimethylamine group being coupled over a methylene group in α-position to an electron-rich, heterocyclic, 5-membered ring system (e.g. furan or thiophene) typically form NDMA with very high conversion rates (Figure 4.13).

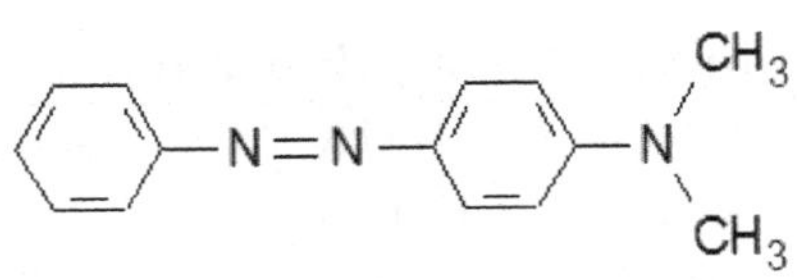 4-(dimethylamino)azobenzene	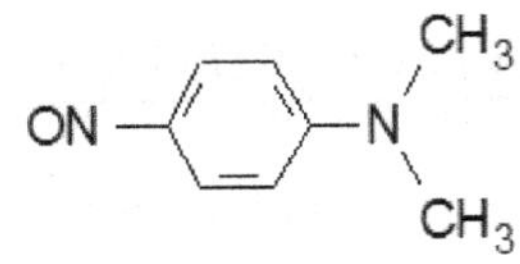 N,N-dimethyl-4-nitrosoaniline
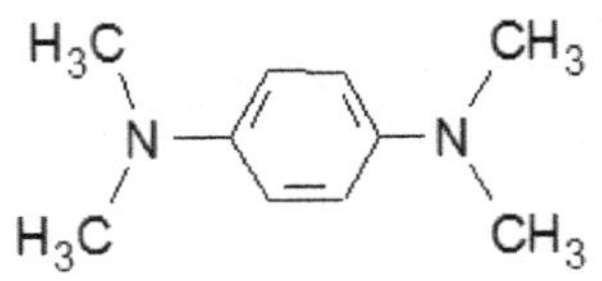 N,N,N‘,N‘-tetramethyl-p-phenylenediamine	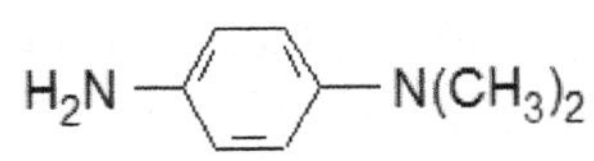 N,N-dimethyl-p-phenylenediamine
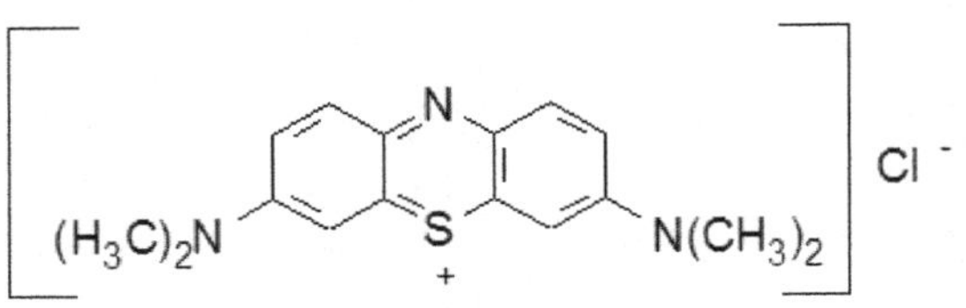 methylene blue	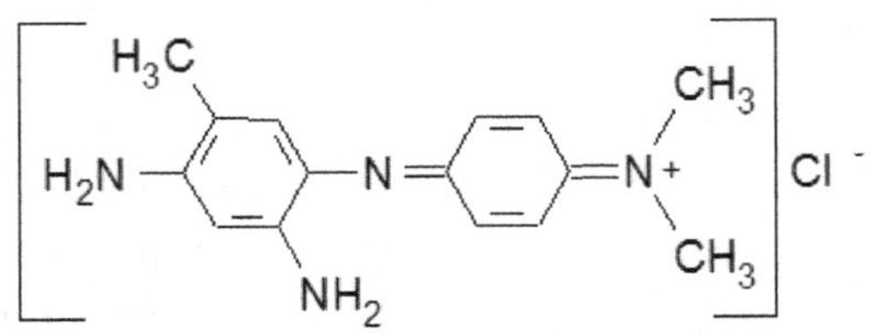 toluylene blue
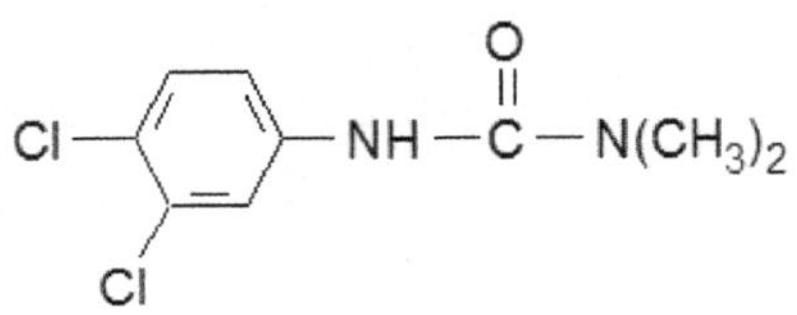 diuron	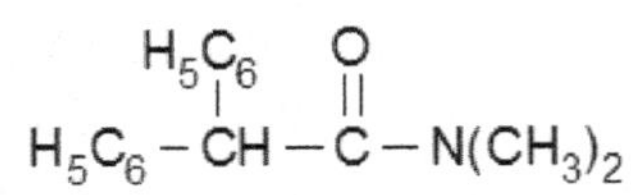 diphenamid
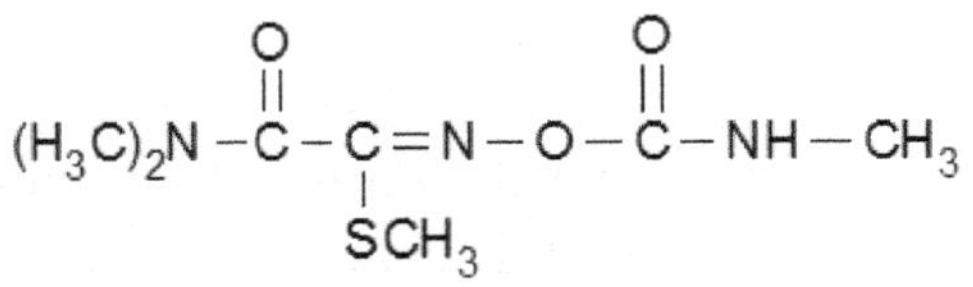 oxamyl	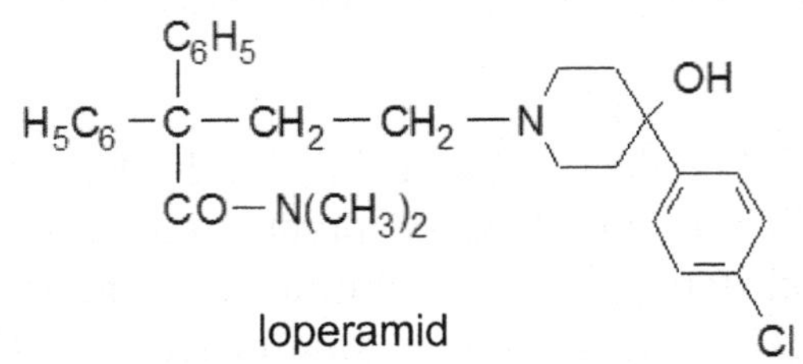 loperamid
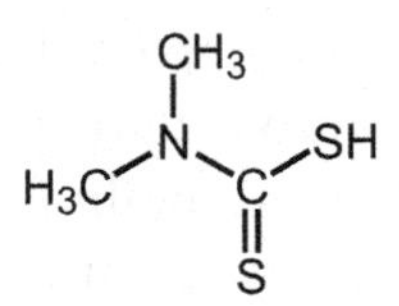 dimethyldithiocarbamate	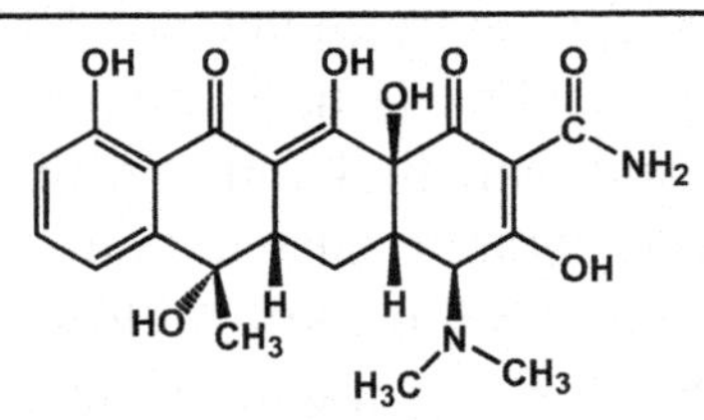 tetracycline

Figure 4.11 Molecular structures of tertiary amines investigated in the NFP test (I)

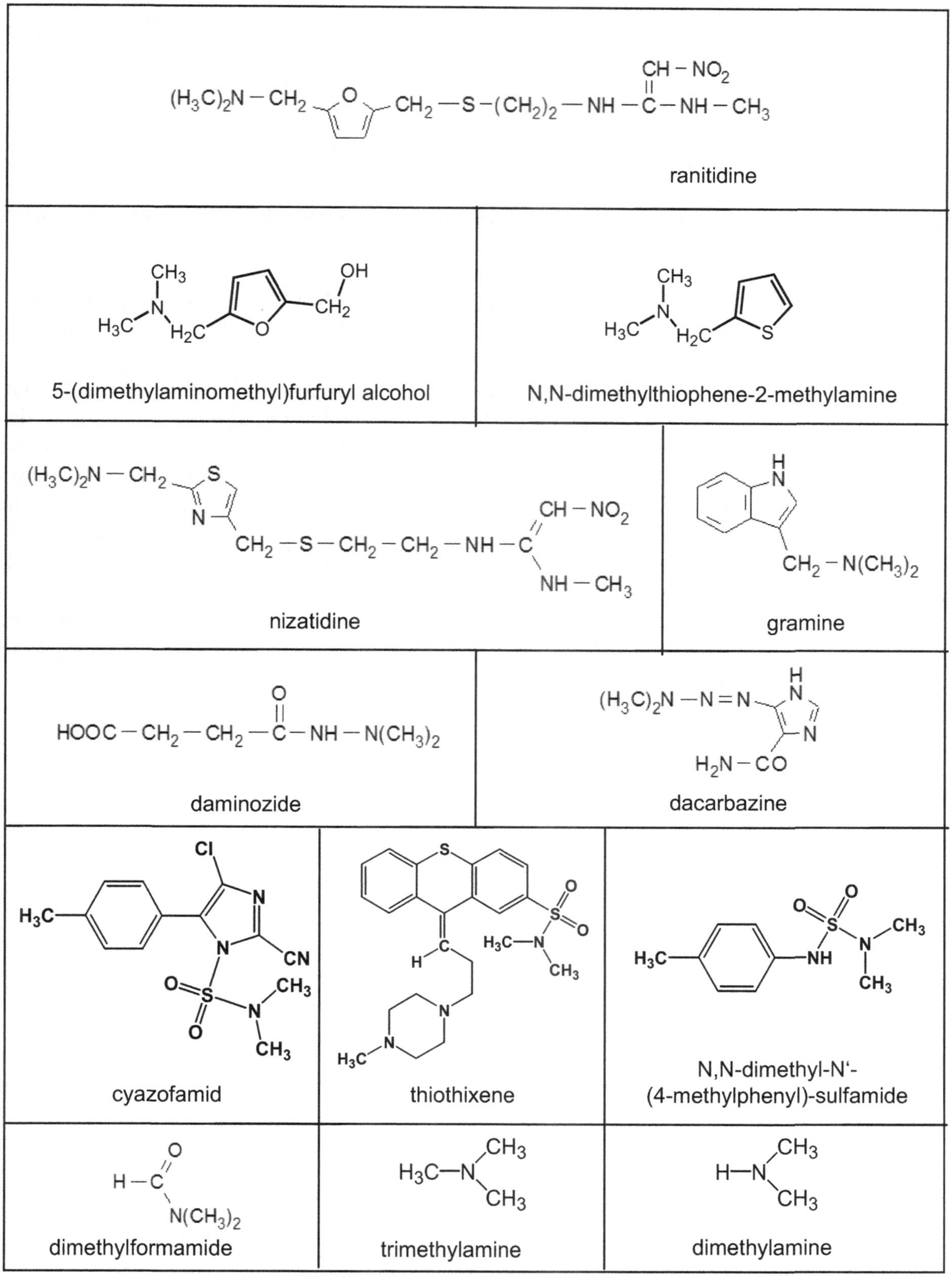

Figure 4.12 Molecular structures of tertiary amines investigated in the NFP test (II)

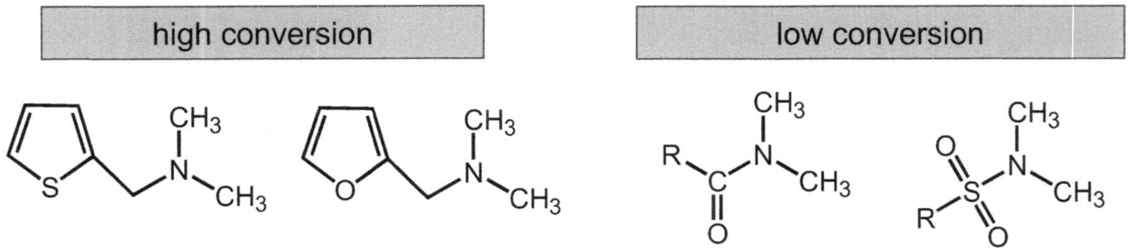

Figure 4.13 Structural characteristics of tertiary dimethylamines facilitating or blocking NDMA formation during chloramination

According to the results presented in this section, NDMA is the major nitrosamine by-product produced during chloramination of surface and groundwaters. Other nitrosamines formed were NDEA, NPYR, and NMOR; however, levels were at least one order of magnitude lower than those of NDMA. Only a small fraction of the formed amounts of NDMA could be attributed to dimethylamine, indicating that the major portion of nitrosamine formation is related to tertiary amines. Based on studies with anthropogenically non-influenced surface waters, the precursor portion within the NOM seems to be rather low. For surface waters, a good correlation was found for NDMA formation and the boron content in the water, supposing a major impact of anthropogenic pollution on the NFP. Conversion efficiencies of tertiary amines were shown to differ substantially, with about 63% of the pharmaceutical ranitidine being transformed to NDMA during chloramination, making it a good candidate for wastewater-related NDMA precursors. The comprehensive evaluation of the data implies that water utilities using surface water under the influence of wastewater are at a higher risk for nitrosamine formation during chloramination than those making use of non-impacted water sources.

CHAPTER 5
REMOVAL OF POSSIBLE NITROSAMINE PRECURSORS

LITERATURE REVIEW

The removal of possible nitrosamine precursors has been studied in three ways: Direct analysis of key amines, analysis of DON levels, and determination of the nitrosamine formation potential (NFP). The majority of investigations, however, were performed with focus on distinct amines and the DON. Only very few studies have yet been performed on the removability of the complete nitrosamine precursor pool by single treatment steps and, so far, no investigation looked at the precursor fate across full-scale drinking water treatment systems.

Microbial Degradation

Between 2 and 70% of the DON in surface waters is reported to be bioavailable, with the high variability being most probably related to differences in its composition (Pehlivanoglu-Mantas and Sedlak 2006). Wastewater derived NDMA precursors measured via the NFP turned out to be relatively stable with respect to biotransformation as indicated by several day half-lives in effluent-dominated surface waters (0-30% removal) or within an engineered treatment wetland receiving nitrified municipal wastewater effluents (ca. 50% removal) (Pehlinvanoglu-Mantas and Sedlak 2006b; Schreiber and Mitch 2006a). Furthermore, NDMA precursor concentrations in effluent samples from municipal wastewater treatment plants were only slightly lowered by 23% during a 30-day incubation experiment in the presence of inocula derived from activated sludge (Pehlinvanoglu-Mantas and Sedlak 2006b). These experiments demonstrate that NDMA precursors released by wastewater treatment plants can persist long enough to possibly impact downstream drinking water treatment plants. The amines involved in the major part of NDMA formation are still unknown. Among amines, however, primary and secondary amines are typically better degradable by microorganisms than tertiary and quaternary amines. The readily biodegradable trimethylamine represents an exception. Alicyclic amines with ring systems possessing one or two heteroatoms are often biologically more stable (Börnick and Schmidt 2006).

Various microorganisms can grow aerobically and anaerobically on TMA, DMA and MA as the sole source of carbon and energy, and the metabolism of these compounds has been elucidated (Colby and Zatman 1973; Boulton and Large 1977; Meiberg and Harder 1978). The initial catabolism of TMA in methylotrophic bacteria may proceed by one of two different pathways. The first route involves the direct demethylation to DMA and formaldehyde (Colby and Zatman 1971), while for the other pathway the initial attack is an oxygenation of TMA to trimethyl-N-oxide (Large et al. 1972). The N-oxide is subsequently demethylated to DMA and formaldehyde (Large 1971; Myers and Zatman 1971). DMA is oxidized to MA and formaldehyde (Eady and Large 1969; Eady et al. 1971). MA may be further oxidized by microorganisms (Eady and Large 1968, 1971; Bamforth and Large 1977).

According to respirometric assays by Calamari et al. (1980) aliphatic amines with linear alkyl chains are easily degradable by bacteria. The authors investigated the biodegradability of DMA, DEA, and DBA by bacteria originating from river mud, sewage sludge, and adapted bacteria from a culture having the amine under investigation as the only carbon source for 15 days. DMA was better degradable than DEA and DBA. Although amines underwent a complete primary degradation, as it was shown by analytical disappearance, oxygen consumption lower than the theoretical one indicated that mineralization was incomplete.

In soils that were incubated at field capacity, DMA quickly disappeared within 14 days (Ayanaba et al. 1973), presumably because it was microbiologically dealkylated. It was also shown that DMA is a metabolite of TMA in soils. Microbial soil populations readily converted TMA to DMA at pH values of 5.8 and 6.5, but this conversion was slow in the soil at pH 3.8. The amount of TMA accumulating as DMA in the soils was generally small with maximum 3%. The failure of DMA to accumulate in higher amounts was attributed to its excellent biodegradability (Ayanaba et al. 1973). Also Tate and Alexander (1976a) reported that DMA was produced from TMA in soils at different pH values (pH 4.5-7.0). The rate of DMA disappearance from the soil was affected by both pH and redox milieu with DMA removal rates being slower under anaerobic conditions. However, no DMA was detectable after 21 days in any of the soils tested.

Islam et al. (1998) studied the fate of dissolved odorous compounds including DMA and TMA in sewage treatment plants and found that concentrations of both amines were substantially lowered. The elimination of TMA turned out to be somewhat lower than that of DMA. Boethling and Alexander (1979) investigated the biodegradation of DMA, DEA and DELA in surface water under aerobic conditions. Incubations at 29 °C revealed no major differences between the two amines. The maximum rate of biodegradation was proportional to the initial amine concentration over a range of concentrations from several nanograms to several milligrams per liter. At the lowest initial level of 21 ng/L DELA, degradation continued at an approximately constant rate for at least 4 days. At this time about 30% of the original compound had been converted to CO_2.

Pitter and Chudoba (1990) published data on the biodegradability of various organic nitrogen compounds determined under model conditions (model water, acclimated cultures, and concentrations in the mg/L range). Using the ratio between biological oxygen demand and total oxygen demand, the authors classified the biodegradability of the compounds into readily degradable (NPA, NBA) and rapidly degradable (MA, DMA, EA, DEA, ELA).

In a study by Pietsch et al. (2001), the microbial degradation of DMA, DEA, pyrrolidine, piperidine, and morpholine was investigated under conditions close to those applied during drinking water treatment (concentrations in the low µg/L range). The investigation was performed under aerobic conditions at 20 °C on the basis of spiked bank filtrate that was passed over a test filter with a growth carrier loaded with mixed bacteria culture from Elbe river. DMA, piperidine, and pyrrolidine were much better removable than DEA and morpholine. In an aerated biofilter studied by Hwang et al. (1994), TMA, DMA and NPA were efficiently removed under both aerobic and anaerobic conditions. Among these amines, however, DMA required the longest acclimation period.

Degradation rates and pathways of MA, DMA and TMA were also investigated in anoxic salt marsh sediments by Wang and Lee (1995). Radio-labeled amines added to the sediments were partly adsorbed, but predominantly metabolized within 10-20 hours. Experiments using specific inhibitors and labeled substrates demonstrated that the decomposition of amines was mediated by methanogens, sulfate reducers, and other microbial groups.

Tetramethylammonium salts were shown to be degradable by whole cells and cell-free extracts of bacteria isolated from sewage (Hampton and Zatman 1973; Mackrell and Walker 1978). Metabolic intermediates were formaldehyde and ammonia. Investigations at wastewater treatment plants demonstrated that dimethylformamide (DMFA) is readily eliminated during secondary treatment (Landesumweltamt Brandenburg 2003).

Coagulation, Flocculation and Sedimentation

Randtke and McCarty (1979) investigated the removal of DON by chemical precipitation. Ferric chloride, lime, and aluminum were less effective than activated carbon, but more effective than ion exchange at neutral pH. Mean DON removal was 40% for ferric chloride, 33% for lime, and 28% for aluminum. According to Westerhoff et al. (2006), maximum DON removal by coagulation with aluminum salts was about 40%. Coagulation with aluminum salts in conjunction with low dosages of the cationic polymer polyDADMAC improved DON removal by 15-20%. Coagulation preferentially removed higher molecular weight fractions of DON (Westerhoff et al. 2006).

Laboratory-scale studies by Pietsch et al. (2001) showed that aliphatic amines can not be completely removed by flocculation with iron(III) or with aluminum salts. Removal efficiency did not exceed 10%, with the exception of ethanolamine and ethylenediamine showing 30 and 45% removal. The results were explained by the positive charge of the amines which inhibit coagulation with the flocculants.

Ozonation and Chlorination

In laboratory-scale experiments with various aliphatic amines, a rather slow decay was observed after application of ozone. Treatment of amines with ozone has to be considered as rather inefficient (Hwang et al. 1994; Mitch et al. 2003a). When a mixture containing 1.6 mg/L TMA, 3.2 mg/L DMA and 5.2 mg/L NPA was ozonated with 3.5 mg/L ozone, only 45% of TMA and about 20% of DMA and NPA were eliminated after 100 min. Using higher dosages of ozone of 9.6 mg/L and initial amine concentrations of 1.3 mg/L, 2.0 mg/L, and 3.3 mg/L for TMA, DMA, and NPA, respectively, TMA was completely removed after 1 hour and NPA after 4 hours. For DMA, 90% removal was achieved after 12 hours. Nitromethane was formed as a by-product of TMA ozonation.

The rate constant for the reaction of DMA with ozone (at 20°C and pH 5-6 was determined to be 2 x 10^7 $M^{-1}s^{-1}$ for initial DMA concentrations of 0.04-4 mmol/L (Hoigné and Bader 1983). The reaction followed pseudo-first order kinetics. Also Pietsch et al. (2001) investigated the behavior of amines during ozonation. The authors determined reaction rate constants by spiking model waters at pH 7.0 with 1 µg/L of amine, followed by an addition of 1 mg/L ozone. In all experiments, the decline in amine concentration was monitored. Additional experiments with an added radical scavenger (*tert*-butyl alcohol) showed that the decline in amine concentration was mainly due to the direct reaction of ozone with the amine.

Table 5.1 shows the corresponding rate constants determined by Pietsch et al (2001). The rate constant strongly depends on the pH of the solution, since only the free amine reacts with ozone, while the protonated amine is practically non-reactive (Munoz and von Sonntag 2000). The substitution of hydrogen atoms by alkyl groups at the amine-N typically increases the rate constant significantly (von Gunten 2003). Determined values for k_{O3} are relatively low which was mainly attributed to the high pK_A values of DMA, DEA, ELA, cyclohexylamine, pyrrolidine, and piperidine. These compounds are predominantly present in their protonated form at pH 7, whereas morpholine, ethylenediamine and piperazine with lower pK_A values are available in their non-protonated form, resulting in higher reaction rate constants (Pietsch et al. 2001).

Table 5.1
Reaction rate constants with ozone for nitrogen-containing compounds in model water

Compound	k_{O3} ($M^{-1}s^{-1}$)	$t_{1/2}$ (min)
DMA	33.5	16.6
DEA	9.6	57.5
Ethylenediamine	218	2.5
Ethanolamine	26.1	21.3
Cyclohexylamine	36.0	15.4
Pyrrolidine	23.2	24.1
Piperidine	36.2	15.4
Morpholine	2.9×10^3	0.19
Piperazine	4.0×10^3	0.14

Source: Pietsch et al. (2001)

Ozonation products of primary amines are hydroxylamines, followed by oximes and amides which are finally degraded to aldehydes and carboxylic acids. Secondary amines follow the same reaction pathway with the exception of oximes, while tertiary amines degrade to aldehydes and carboxylic acids via N-containing oxides formation (Elmghari-Tabib et al. 1982). In the case of simple tertiary amines, to a lesser extent also the corresponding secondary amines are formed (Munoz and von Sonntag 2000).

Chlorine was shown to react quite fast with aliphatic amines forming organic chloramines (Abia et al. 1998). The rate constant for chlorine transfer from hypochlorous acid to DMA, for instance, was determined to 6.1×10^7 $M^{-1}s^{-1}$. Meanwhile, several reports showed that pre-chlorination of waters can lower the NDMA formation potential during chloramination by 70-90% (Mitch and Sedlak 2002b; Wilczak et al. 2003; Chen and Valentine 2005; Schreiber and Mitch 2005; Mitch et al. 2005; Pehlinvanoglu-Mantas and Sedlak 2006b), indicating that chlorination prior to chloramination may be one strategy to minimize NDMA formation.

Adsorption on Activated Carbon

Studies investigating DON removal by activated carbon (AC) showed that more than half of the wastewater-derived DON (56-76%) could be removed by contact with 15-50 g/L adsorbent (Randtke and McCarty 1979, Keller et al. 1978, Parkin and McCarty 1981). Westerhoff et al. (2006) found a DON removal of 20% from surface water and of 50% from a finished drinking water applying an AC dosage of 5 mg/L. Higher AC dosages of 25 mg/L increased the DON removal only slightly by about 5-10%. Differences in DON removal among AC brands were negligible. Also the adsorption capacity for individual amines was found to be rather low. Hwang et al. (1994) investigated GAC adsorption for the removal of TMA, DMA and NPA. Freundlich coefficients, determined on the basis of isotherms, are shown in Table 5.2. NPA exhibits the highest removal efficiency, even though adsorption capabilities are low in general. Determination of a breakthrough curve for TMA in laboratory-scale column tests showed an immediate breakthrough of 50% which increased rapidly to 100%. Also other sources classify aliphatic amines like NPA and NBA as poorly absorbable on AC (Landesumweltamt Brandenburg 2003).

Table 5.2
Freundlich parameters for precursor adsorption onto GAC

Compound	Initial concentration c_0 (mg/L)	K [GWC / F-400] (µg/g)	1/n [GWC / F-400]
TMA	6	1.06 / 0.84	0.22 / 0.31
DMA	18	0.02 / 7.73	2.64 / 0.26
NPA	18	7.26 / 12.3	0.83 / 0.92

GWC: Granular wood charcoal; F-400: Filtrasorb 400
Source: Hwang et al. (1994)

Strong anion and cation exchangers were shown to be capable of removing about 12-32% and 11-49% of wastewater-derived DON, respectively. Removal efficiencies were pH-dependent, with more removal observed for cation exchangers at lower pH values (Pehlivanoglu-Mantas and Sedlak 2006a).

Alternative Treatment Technologies

Most of the nitrosamine precursors are non-reactive towards UV irradiation due to lack of chromophores in the molecule to be attacked (Mitch et al. 2003a). Since many precursors for nitrosamine formation are charged at neutral pH, aeration experiments should also show no removal of aliphatic amines (Mitch et al. 2003a). On the other hand, particle-associated NDMA precursors can be removed by microfiltration or reverse osmosis treatment. The latter is also capable of removing charged, dissolved NDMA precursors such as protonated dimethylamine (Mitch and Sedlak 2004).

LABORATORY-SCALE EXPERIMENTS

Microbial Degradation

The objective of this part was to investigate the influence of biofiltration processes on the removal of nitrosamine precursor compounds present in wastewater-impacted surface waters. For this purpose, microbial degradation of nitrosamine precursor compounds was studied in the laboratory using a biologically active test filter. In these experiments, three different surface waters were percolated through the bioactive fixed bed for about one month and the decrease in concentration was determined by regularly taking samples from the reservoir after defined time intervals and determination of the precursor amount via the NFP test. This simulation method gives basic information about the biodegradability of NDMA precursor compounds excluding sorption effects to sediment or soil. Investigated surface waters were of medium to high wastewater impact and showed a relatively high NDMA formation potential. Surface waters I and II were taken from the same location, but at different times, and differed with regard to the amount of NDMA formed during chloramination. The fourfold higher NDMA formation potential at the first sampling date was caused by an abnormal occurrence at an industrial wastewater treatment plant (WWTP) upstream of the sampling point due to a full load situation with release of certain amounts of untreated wastewater. Surface water III was taken from another river that is impacted mainly by municipal wastewater.

The degradation behavior of the NDMA precursor compounds occurring in these surface waters is shown in Figure 5.1. It can be seen that NDMA precursor compounds tend to be rather persistent, as concentrations decreased only slowly with roughly 70-90% of the initial amount persisting even after one month runtime. Most interestingly, no major differences in degradation rates were noticed between the different surface waters, even though NDMA precursor compounds were of different initial concentration and origin.

These test filter results give an impression of the microbial degradability of precursor compounds relevant for NDMA formation during chloramination and leads to the hypothesis that their removal by biofiltration processes like riverbank filtration or artificial groundwater recharge might be low. In these processes, however, sorption effects that were intentionally excluded in the test filter experiment can also be relevant. During subsurface transport, the filtrate is in contact with the solid aquifer material and adsorption onto solid material might take place.

It has to be taken into account that the bulk of NDMA precursor compounds certainly consists of more than one substance. Thus, certain precursors like DMA might be degradable while others are not attacked by microorganisms. The present experiments demonstrate that the major precursor part is certainly rather persistent towards biodegradation and confirm more recent experiments by Pehlinvanoglu-Mantas and Sedlak (2006b) and Schreiber and Mitch (2006a), reporting on the limited biotransformation of NDMA precursors in effluent-dominated surface waters.

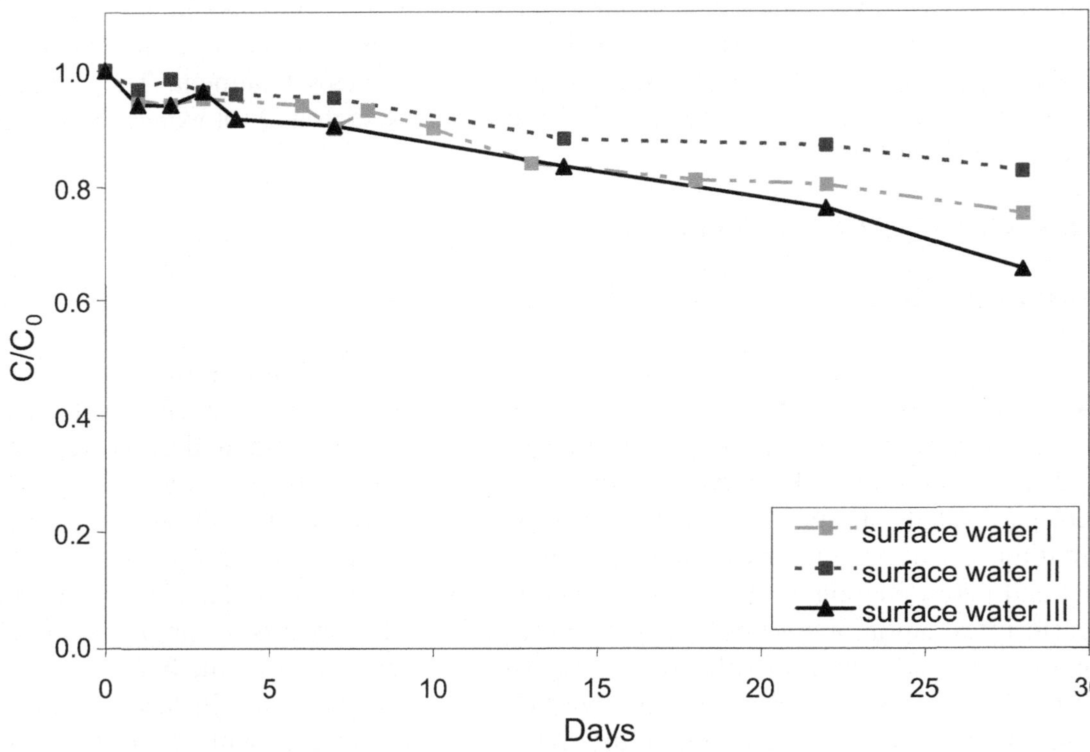

Conditions: NFP test (0.4 mM chloramine, pH 7, 20 °C, 168 h), c_0 surface water I – 1000 ng/L NDMA, c_0 surface water II – 270 ng/L NDMA, c_0 surface water III – 84 ng/L NDMA

Figure 5.1 Microbial degradation of NDMA precursor compounds in a test filter based on measurements by the NFP test

Coagulation, Flocculation and Sedimentation

The impact of a coagulation, flocculation and sedimentation process on the removal of nitrosamine precursor compounds was investigated in laboratory-scale tests as described in Chapter 2. Four German river waters of different origin and with different impact of municipal and industrial wastewater were investigated. The results of the NFP tests performed after the flocculation experiments with different flocculants (iron chloride, $FeCl_3$ and polyaluminum chloride, $pAlCl_3$) and different flocculant doses (5 and 10 mg/L) are presented in Figures 5.2 to 5.5 for NDMA, in Figure 5.6 for NDEA and in Figures 5.7 to 5.10 for NPYR. NDEA was only found after chloramination of one river water. Other nitrosamines than NDMA, NDEA and NPYR could not be detected, neither in the original water samples nor after the flocculation experiments.

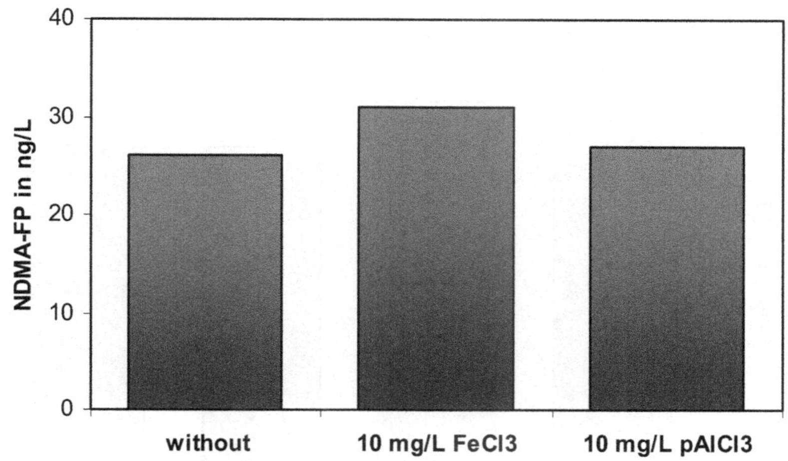

Conditions: 0.4 mM chloramine NFP test, 20 °C, pH = 7.0, 168 h

Figure 5.2 Impact of coagulation, flocculation and sedimentation on NDMA formation during chloramination of river water I

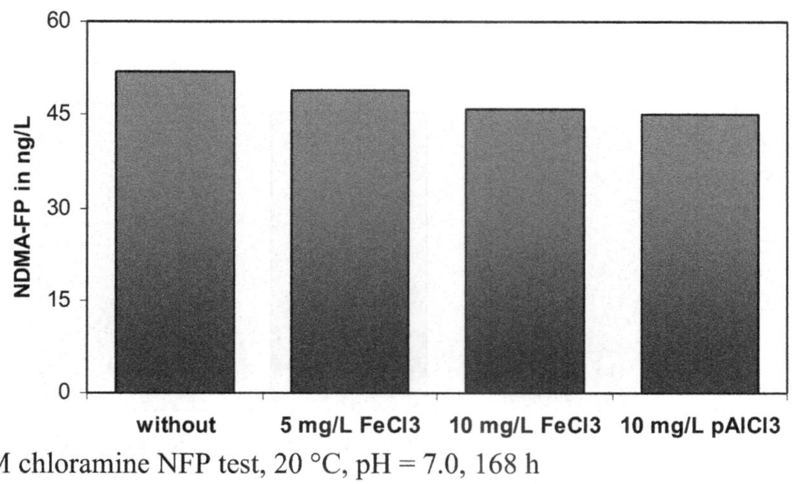

Conditions: 0.4 mM chloramine NFP test, 20 °C, pH = 7.0, 168 h

Figure 5.3 Impact of coagulation, flocculation and sedimentation on NDMA formation during chloramination of river water II

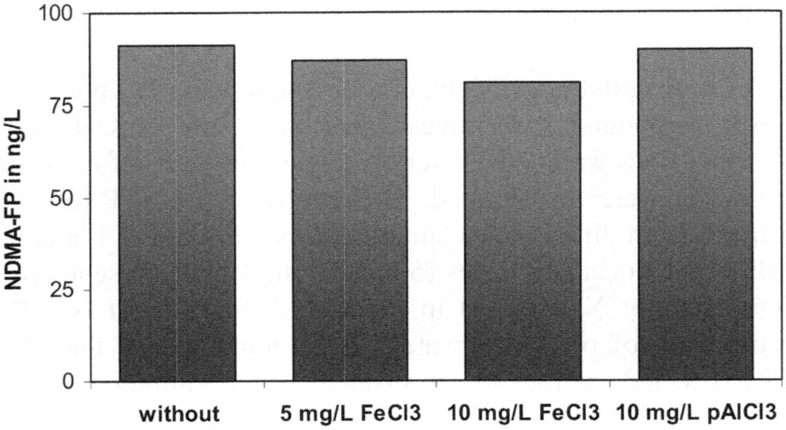

Conditions: 0.4 mM chloramine NFP test, 20 °C, pH = 7.0, 168 h

Figure 5.4 Impact of coagulation, flocculation and sedimentation on NDMA formation during chloramination of river water III

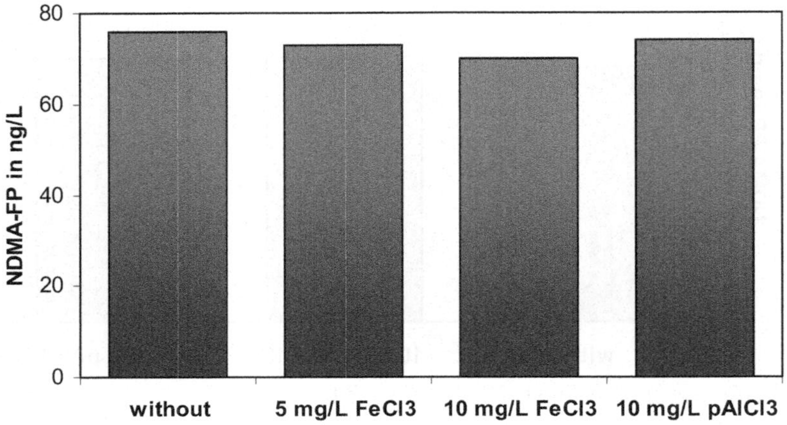

Conditions: 0.4 mM chloramine NFP test, 20 °C, pH = 7.0, 168 h

Figure 5.5 Impact of coagulation, flocculation and sedimentation on NDMA formation during chloramination of river water IV

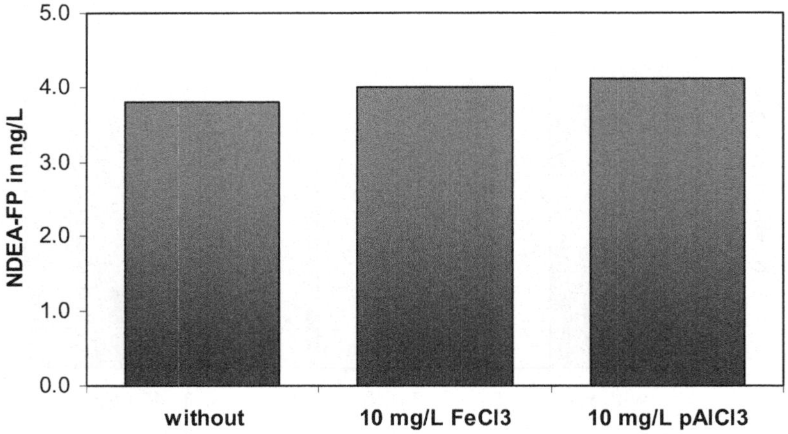

Conditions: 0.4 mM chloramine NFP test, 20 °C, pH = 7.0, 168 h

Figure 5.6 Impact of coagulation, flocculation and sedimentation on NDEA formation during chloramination of river water I

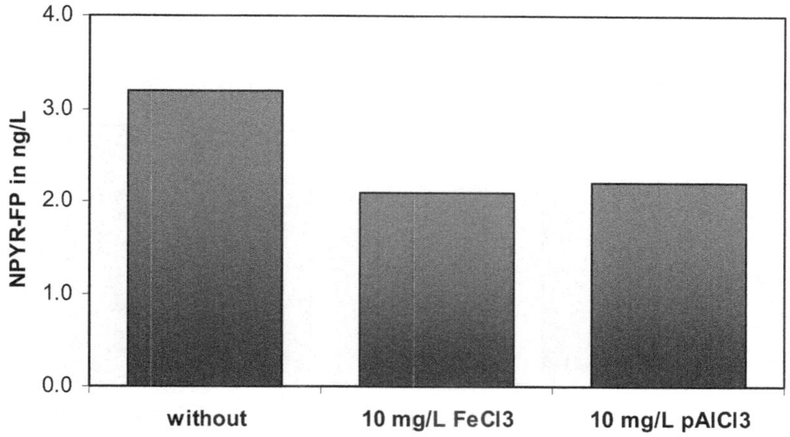

Conditions: 0.4 mM chloramine NFP test, 20 °C, pH = 7.0, 168 h

Figure 5.7 Impact of coagulation, flocculation and sedimentation on NPYR formation during chloramination of river water I

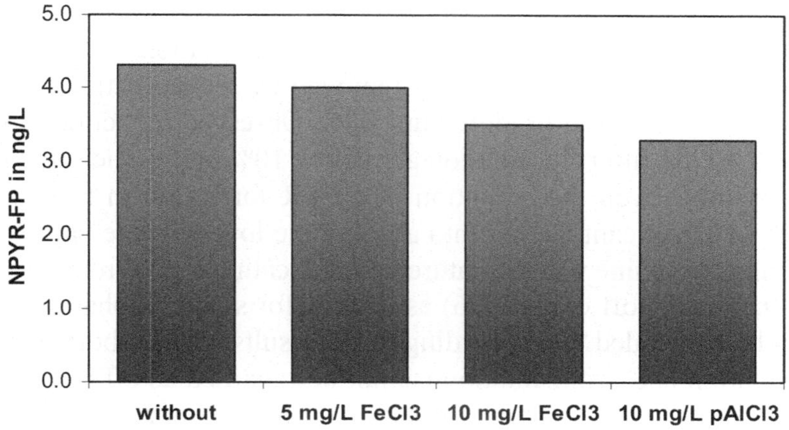

Conditions: 0.4 mM chloramine NFP test, 20 °C, pH = 7.0, 168 h

Figure 5.8 Impact of coagulation, flocculation and sedimentation on NPYR formation during chloramination of river water II

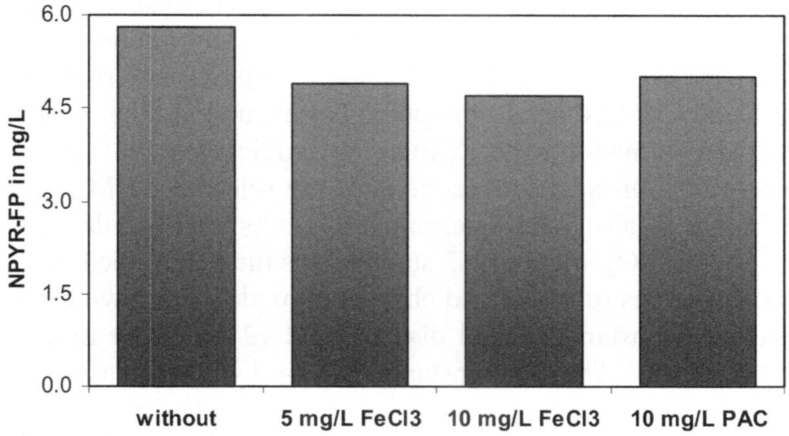

Conditions: 0.4 mM chloramine NFP test, 20 °C, pH = 7.0, 168 h

Figure 5.9 Impact of coagulation, flocculation and sedimentation on NPYR formation during chloramination of river water III

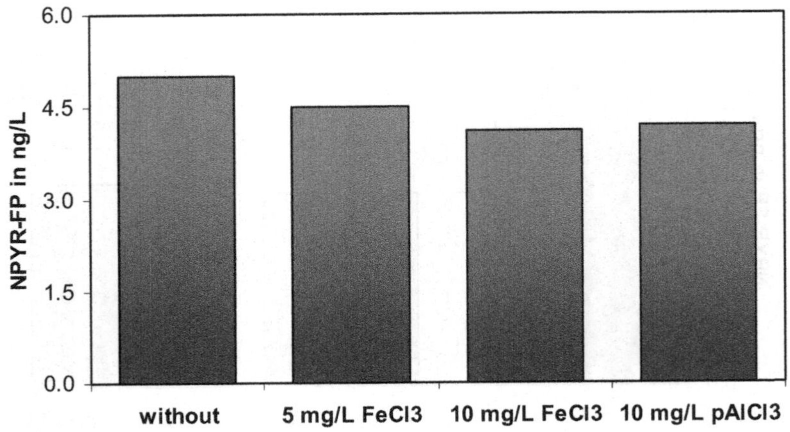

0.4 mM chloramine NFP test, 20 °C, pH = 7.0, 168 h

Figure 5.10 Impact of coagulation, flocculation and sedimentation on NPYR formation during chloramination of river water IV

The results show that for all river waters under investigation the precursor compounds for nitrosamine formation cannot be significantly removed by a coagulation, flocculation and sedimentation process with either iron or aluminum salts. Observed reduction rates for formation of NDMA, NDEA and NPYR after flocculation are below 10% and in the same magnitude than the experimental uncertainty. Even the reduction in NPYR formation in river water I (Figure 5.7) cannot be regarded as significant, taking into account the low absolute value of the NPYR levels formed. These results are in line with literature data indicating a poor removal for DON (Randtke and McCarty 1979; Westerhoff et al. 2006) as well as for single aliphatic amines (Pietsch et al. 2001). Thus it can be concluded that according to the results of the laboratory-scale experiments, coagulation, flocculation and sedimentation cannot be regarded as efficient treatment option for the removal of precursors for nitrosamine formation during disinfection.

Oxidation

Potential NDMA precursors for oxidation experiments were selected from previous studies (see Figure 5.11). Dimethylamine (DMA) and five tertiary amines (TMA, DMEA, DMFA, DMTC, and DMAB) are potential NDMA precursors found in municipal wastewater (Mitch and Sedlak 2004). DMAI and DMAP are from malted barley, and are the responsible tertiary amine alkaloids for the NDMA formation in beer (Mangino and Scanlan 1985; Poocharoen et al. 1992).

Second-order rate constants for reactions of the selected NDMA precursors with ozone (O_3) and chlorine dioxide (ClO_2) were determined to assess the degradation kinetics during ozonation and oxidation with ClO_2. Figure 5.12 summarizes the determined rate constants as a function of solution pH. Reactions of ozone and chlorine dioxide are known to be highly selective so that the determined rate constants ranged over 10 and 12 orders of magnitude for ozone and chlorine dioxide, respectively. The rate constants increased with increasing pH since the deprotonated species of NDMA precursors are stronger electrophiles. DMAB, DMAI and DMAP which all have aromatic ring systems were found to be fast-reacting compounds for both reactions of ozone and chlorine dioxide, whereas NDMA and DMFA reacted very slowly with the oxidants.

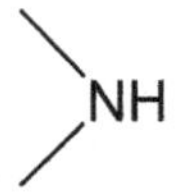

Dimethylamine (DMA)

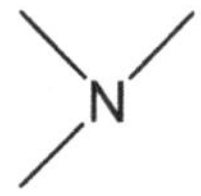

Triimethylamine (TMA)

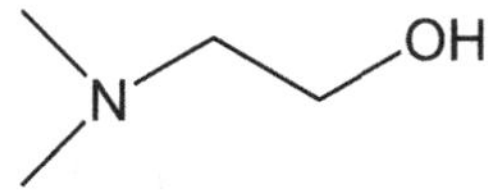

Dimethylethanolamine (DMEA)

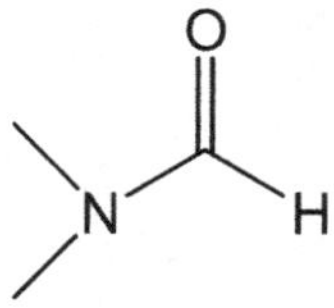

Dimethylformamide (DMFA)

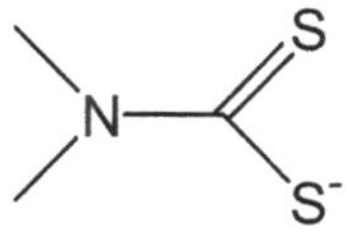

Dimethyldithiocarbamate (DMTC)

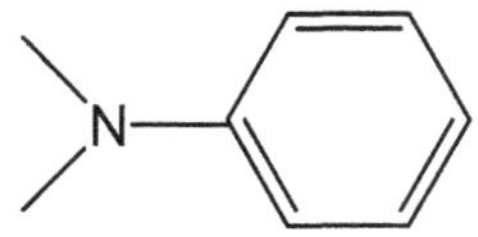

Dimethylaminobenzene (DMAB)

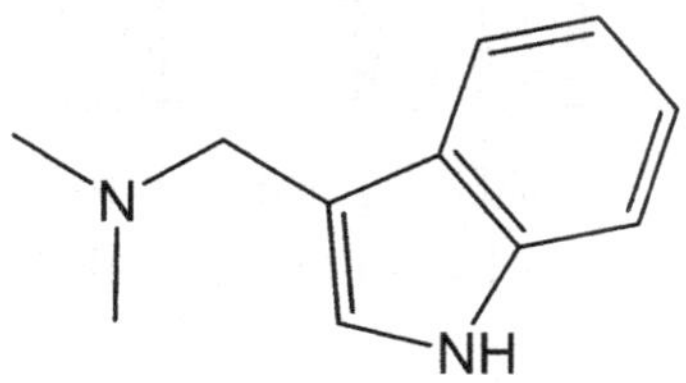

3-(Dimethylaminomethyl)indole (DMAI)

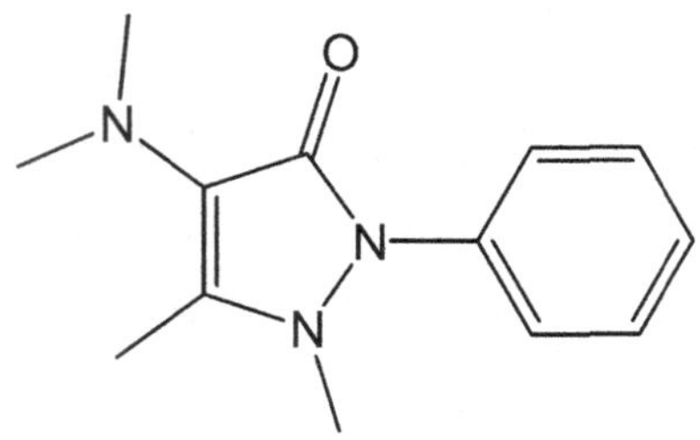

4-Dimethylaminoantipyrine (DMAP)

Figure 5.11 Chemical structures of NDMA precursors selected for oxidation experiments

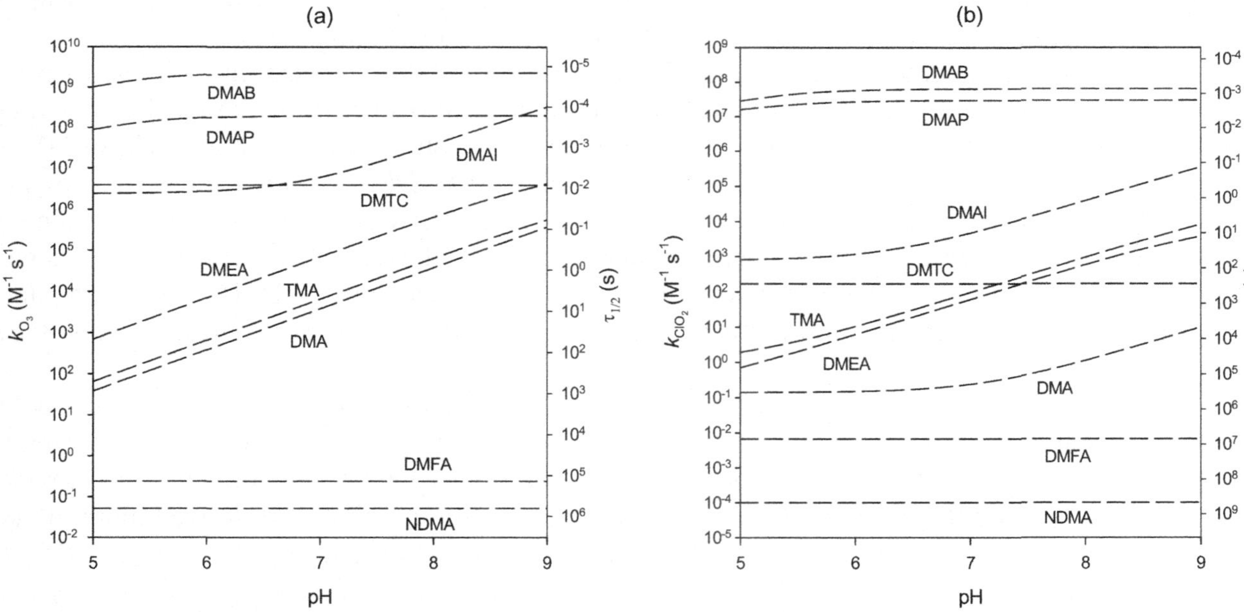

The half-lives are calculated for (a) an ozone concentration of 1 mg/L (20.8 μM) neglecting reactions with hydroxyl radicals, for (b) a chlorine dioxide concentration of 1 mg/L (14.8 μM)

Figure 5.12 Apparent second-order rate constants and half-lives for the reactions of NDMA and its precursors with (a) ozone, (b) chlorine dioxide as a function of pH

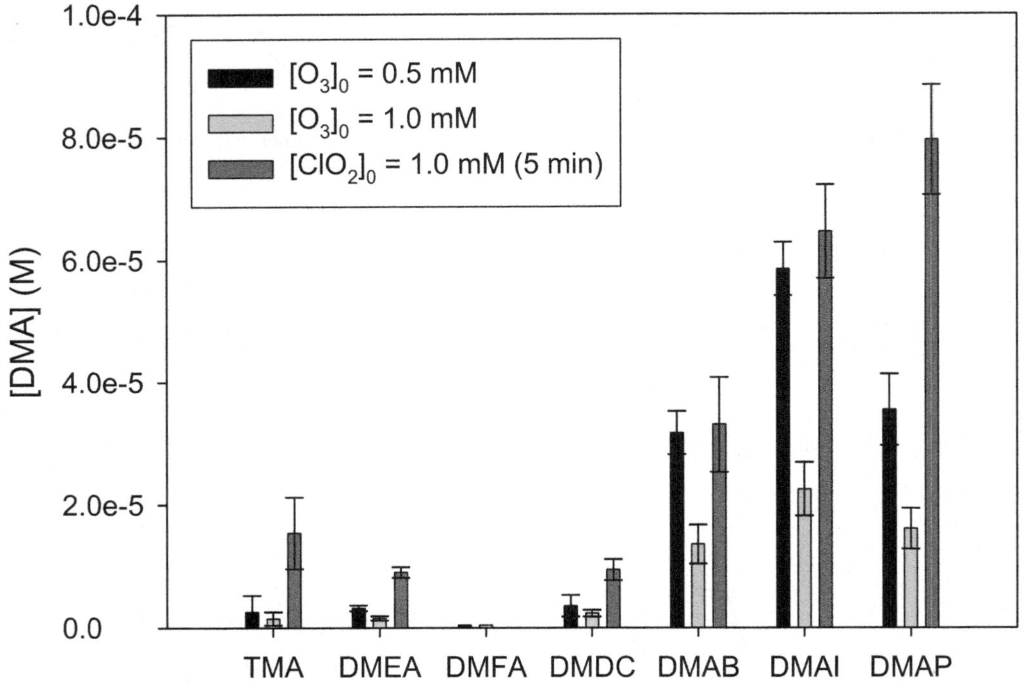

Conditions: $[\text{Precursor}]_0 = 1\times10^{-4}$ M, $[O_3]_0 = 5\times10^{-4}$ M, $[ClO_2]_0 = 1\times10^{-3}$ M (5 min), pH 7, 10 mM phosphate buffer

Figure 5.13 DMA yields from oxidation of NDMA precursors by ozone and chlorine dioxide

DMA is the simplest and the most important NDMA precursor, and is possibly produced from the oxidation of tertiary amine with DMA functional group. DMA yields were measured after the complete oxidation of tertiary amine NDMA precursors (Figure 5.13).

Looking at the data in Figure 5.13, the oxidation of NDMA precursors with chlorine dioxide tends to yield a higher concentration of DMA than with ozone. In particular, fast-reacting compounds as DMAB, DMAI, and DMAP (see Figure 5.12) showed a very high yield of DMA from the oxidation by both ozone and chlorine dioxide. The DMA yields decreased with increasing ozone dose from 0.5 mM to 1 mM (Figure 5.13), whereas no significant reduction of DMA yields was observed with a higher exposure of chlorine dioxide (data not shown). This behavior is ascribed to the difference of the rate constants for reactions of DMA with ozone and chlorine dioxide, which are approximately 10^3 and 10^{-1} $M^{-1}s^{-1}$ at pH 7, respectively (see Figure 5.12).

Figure 5.14 shows the NDMA formation potentials after oxidation of the precursors with ozone and chlorine dioxide. Interestingly, DMAI and DMAP show a higher NDMA FP than DMA. For all compounds, the oxidation of NDMA precursors with ozone and chlorine dioxide significantly reduced the NDMA FP. In particular, the oxidation with ozone showed more effective reduction of NDMA formation potential than that with chlorine dioxide except for DMEA oxidation. The trend of NDMA FP of oxidized precursors can be effectively explained by the DMA yields presented in Figure 5.13. In most cases, a higher DMA yield leads to a higher NDMA FP.

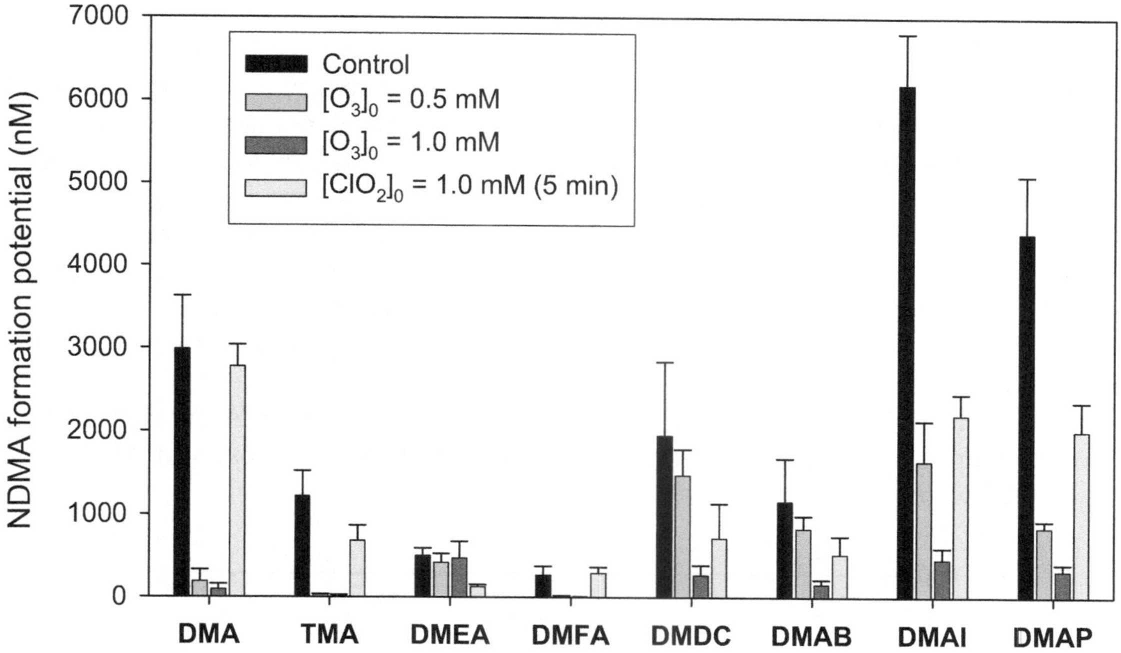

Conditions: $[\text{Precursor}]_0 = 1\times10^{-4}$ M, pH 7, 10 mM phosphate buffer

Figure 5.14 Effect of treatment by ozone and chlorine dioxide on NDMA formation potential

To investigate the effect of pre-oxidation on the NDMA formation potential of natural waters, experiments were performed using water samples from three rivers and one lake. Table 5.3 shows NDMA FP values of the natural waters under investigation after pre-oxidation using various oxidants. Iron(VI), known as ferrate, is a powerful oxidant and its reactions with pollutants are typically fast with the formation of non-toxic by-products (Sharma et al. 2005). As expected from the results with model NDMA precursors, the pre-oxidation significantly reduced the NDMA FP of natural waters. However, the effectiveness of chlorine dioxide and chlorine for reduction of NDMA FP was found to be similar to that of ozone, whereas in the experiments with model compounds ozone proved to be more efficient than chlorine dioxide. Further investigations are necessary to explain this behavior, but it may be originated from the difference between the model NDMA precursors selected and the real ones present in the natural waters used.

Table 5.3
Effect of pre-oxidation on NDMA FP of natural waters

Sample	River water 1	River water 2	River water 3	Lake water
Control	370	70	160	44
20 μM O_3	100	33	52	30
40 μM O_3	28	34	44	30
40 μM ClO_2	24	28	55	23
40 μM HOCl	23	30	39	19
20 μM Fe(VI)	350	55	120	38
0.5 mM Fe(VI)	58	38	42	110

NDMA formation potential in ng/L

Adsorption

For studying the removal of precursor compounds for nitrosamine formation by activated carbon adsorption, laboratory-scale batch experiments with powdered activated carbon (PAC) were performed as described in Chapter 2. For these experiments, several river water samples were selected which exhibited significant differences in their NFP in previous experiments.

The first river water sample under investigation yielded an elevated formation potential for NDMA of more than 700 ng/L. The other river water samples used for the adsorption experiments yielded significantly lower NDMA formation potentials. For river IV, the formation of NMOR and NPYR was also observed and the reduction of the respective formation potential by PAC adsorption was studied as well.

Figures 5.15 to 5.20 summarize the formation potentials for NDMA, NMOR, and NPYR of the water samples under investigation and the changes that were observed when different amounts of PAC were added.

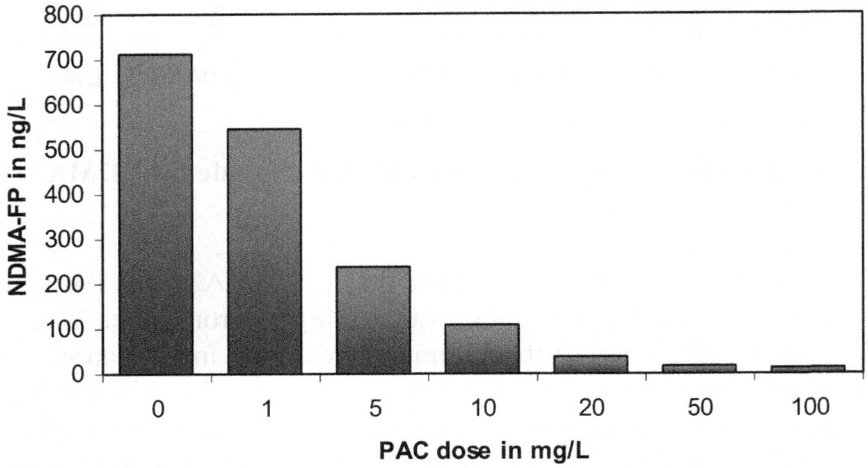

Figure 5.15 Removal of NDMA precursors by powdered activated carbon in river water I

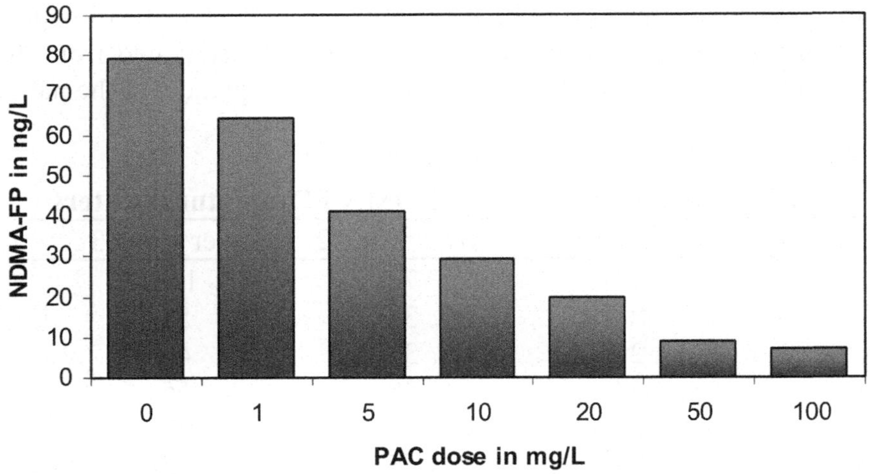

Figure 5.16 Removal of NDMA precursors by powdered activated carbon in river water II

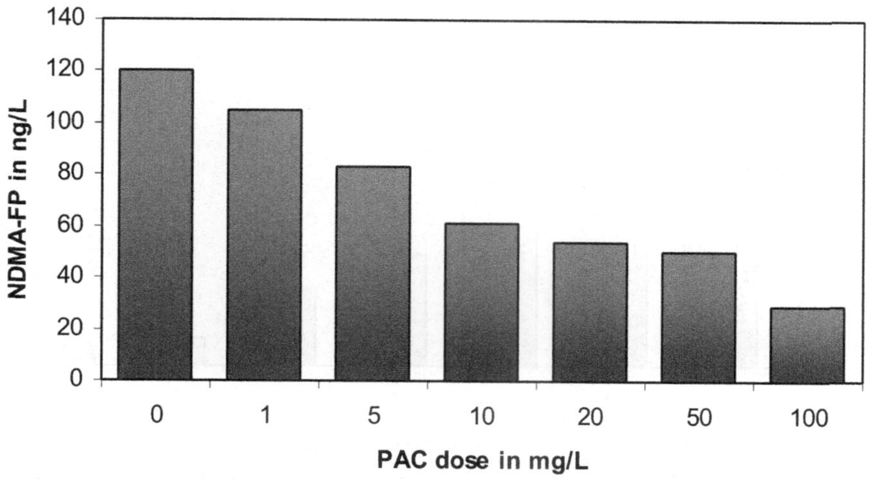

Figure 5.17 Removal of NDMA precursors by powdered activated carbon in river water III

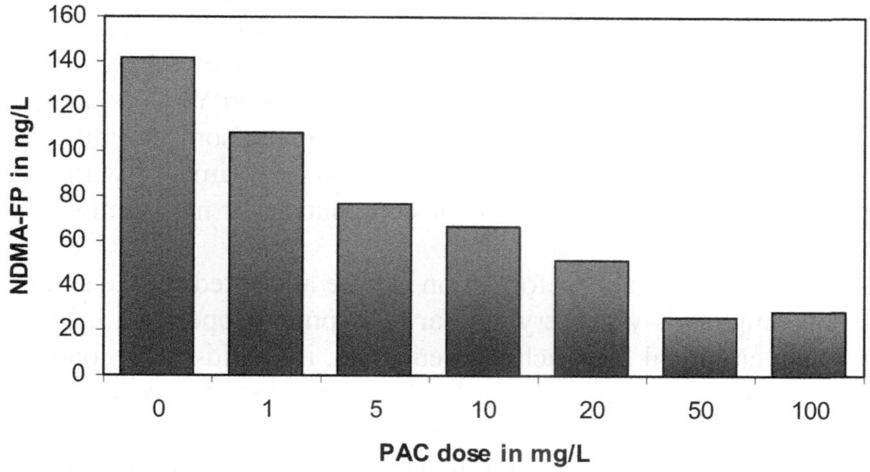

Figure 5.18 Removal of NDMA precursors by powdered activated carbon in river water IV

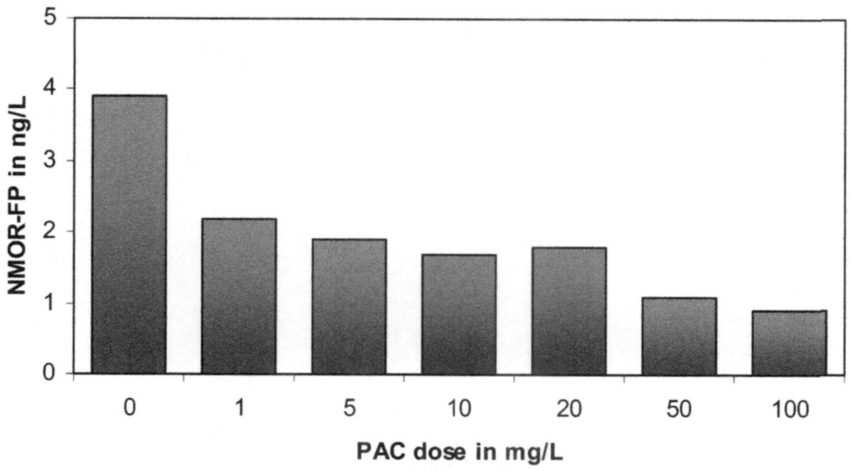

Figure 5.19 Removal of NMOR precursors by powdered activated carbon in river water IV

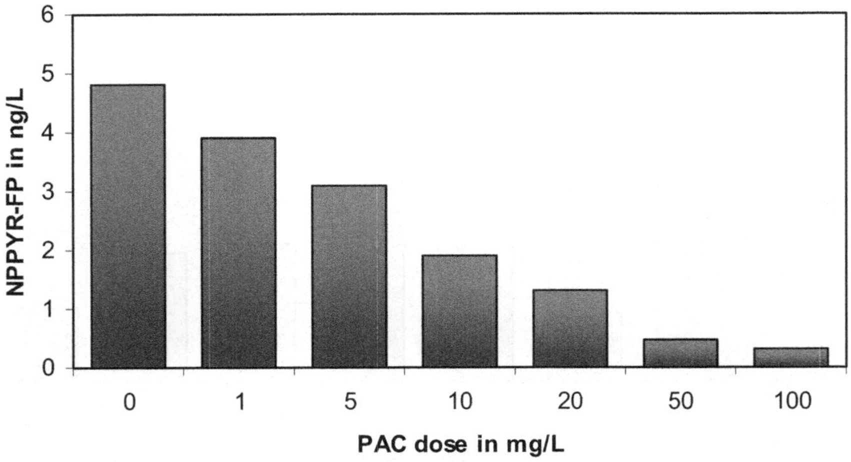

Figure 5.20 Removal of NPYR precursors by powdered activated carbon in river water IV

As can easily be seen from the results presented in Figures 5.15 to 5.20, the amounts of nitrosamines formed in the NFP tests decrease with increasing PAC dose. Already by addition of a PAC dose of 5 mg/L, the NDMA formation potential of the river waters could be decreased by at least a factor of 2 and the same is true for the NMOR and NPYR formation potential in river water IV (Figures 5.19 and 5.20). A qualitative comparison of the data presented in Figures 5.15 to 5.20 with results of similar adsorption experiments with single compounds (Sacher et al. 2001) indicates that the precursor compounds for nitrosamine formation are well adsorbable onto activated carbon.

Assuming that the nitrosamine formation can be attributed to one single precursor compound (or a set of compounds with very similar adsorption properties), data evaluation can be done according to the empirical approach of Freundlich, i.e. solid-phase concentration (here: the part of NFP adsorbed onto the activated carbon) can be plotted in a double-logarithmic scale against residual liquid-phase concentration (here: the residual NFP in the water sample). Figures 5.21 to 5.26 give the resulting Freundlich plots for the river waters under investigation.

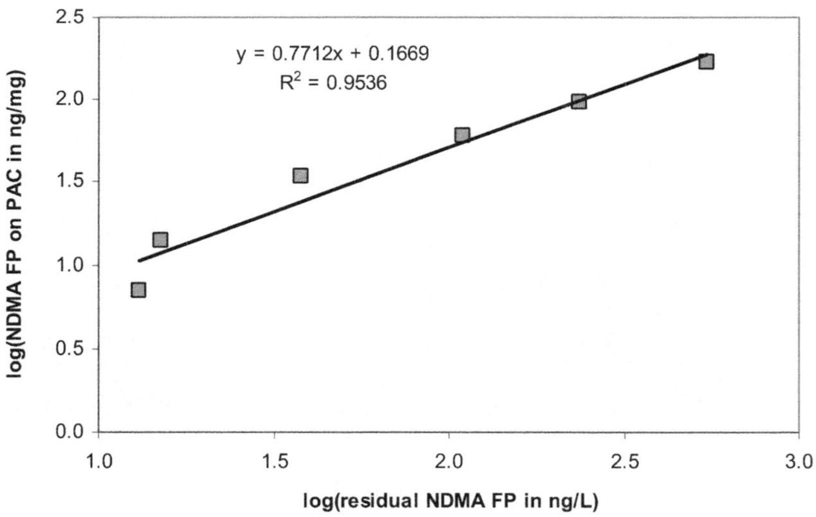

Figure 5.21 Freundlich plot for the NDMA formation potential in river water I

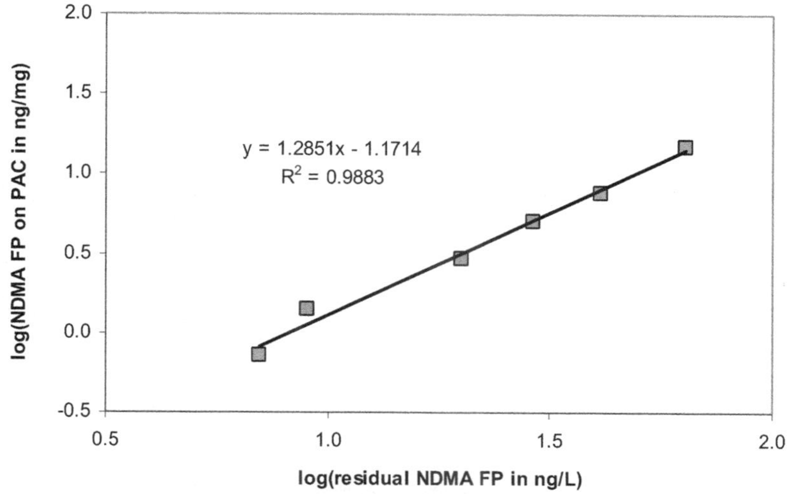

Figure 5.22 Freundlich plot for the NDMA formation potential in river water II

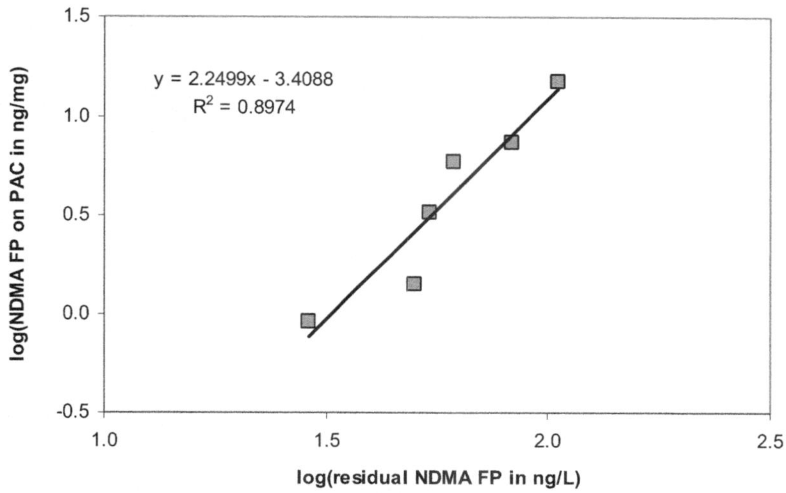

Figure 5.23 Freundlich plot for the NDMA formation potential in river water III

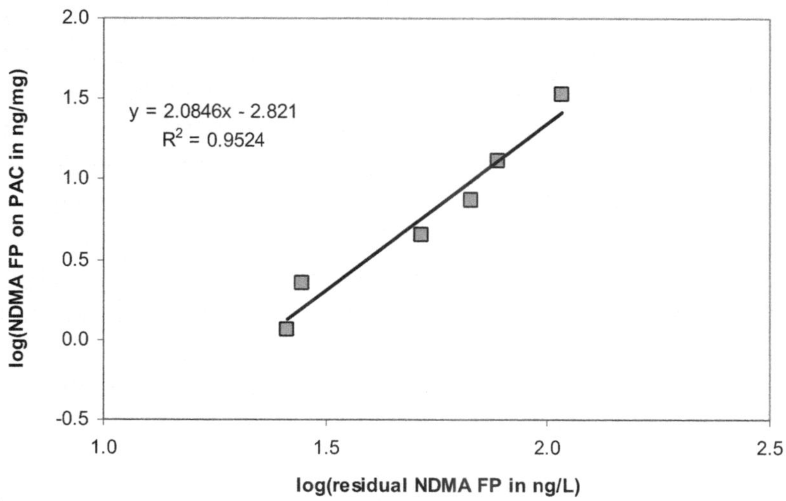

Figure 5.24 Freundlich plot for the NDMA formation potential in river water IV

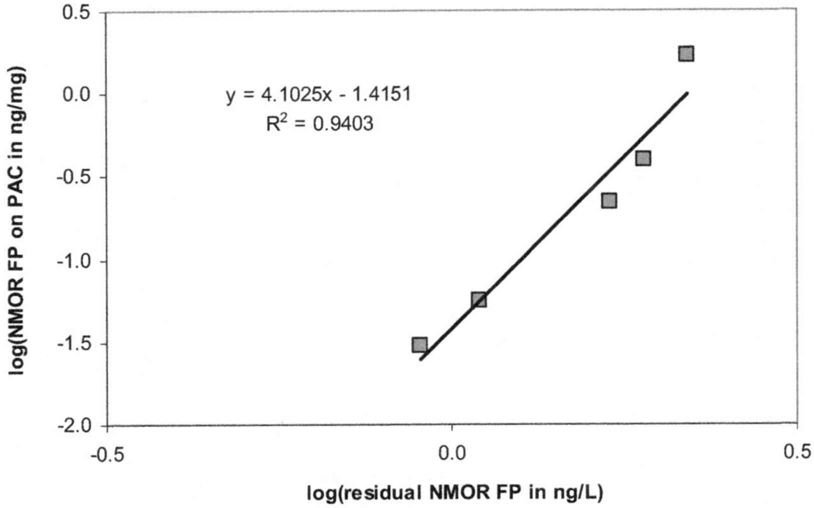

Figure 5.25 Freundlich plot for the NMOR formation potential in river water IV

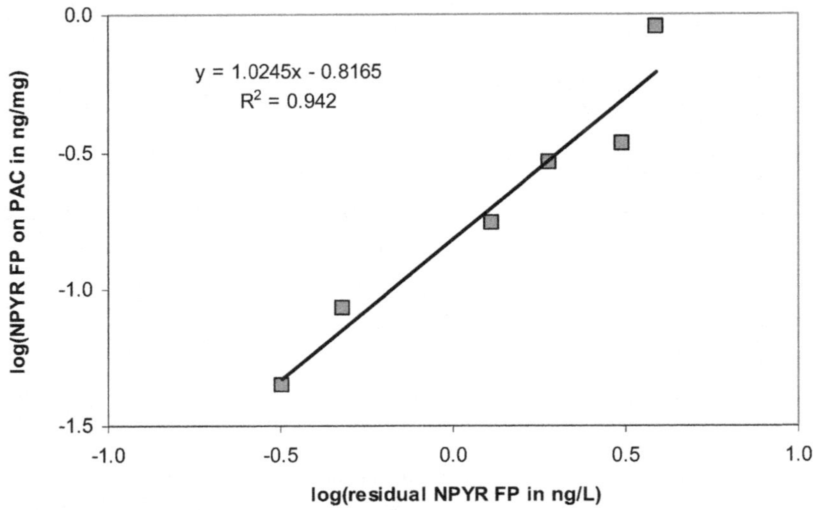

Figure 5.26 Freundlich plot for the NPYR formation potential in river water IV

It can be seen that for all water samples under investigation most of the data points fit quite well to the straight line. Regression coefficients are above 0.9 in most cases indicating a good correlation between the two parameters. Deviations are slightly larger for the NMOR and NPYR formation potential (Figures 5.25 and 5.26) which is mainly due the low absolute amounts of nitrosamine formed. This is a clear indication that NDMA formation in the sample under investigation can be attributed to a single compound or a group of very similar compounds and not to a complex mixture of different compounds. NOM isotherms, e.g. in most cases can not be described by a single Freundlich isotherm as NOM is constituted of a variety of compounds with different adsorption properties. Thus, NDMA formation from NOM (and not from one specific NOM constituent) would result in a different picture.

Table 5.4 summarizes the Freundlich parameters K and 1/n calculated from the intercepts and the slopes of the Freundlich isotherms.

Table 5.4
Freundlich parameters 1/n and K for nitrosamine precursor adsorption onto GAC

Sample	1/n [F300]	K [F300]
NDMA formation potential		
River I	0.77	1.5
River II	1.29	0.067
River III	2.25	0.00039
River IV	2.09	0.0015
NMOR formation potential		
River IV	4.11	0.038
NPYR formation potential		
River IV	1.02	0.15

F300: Filtrasorb 300

The results of the laboratory-scale experiments indicate that the precursors for nitrosamine formation adsorb quite well onto activated carbon. This is in contradiction to the results obtained for small aliphatic amines like DMA which are described in literature as only poorly adsorbable onto activated carbon. Thus it seems as if other compounds than small aliphatic amines play an important role for the formation of nitrosamines during disinfection.

From the results of the laboratory-scale adsorption experiments it can be concluded that adsorption onto activated carbon might be a promising option for the – at least partial – removal of precursor compounds for nitrosamine formation during disinfection.

CHAPTER 6
REMOVAL OF NITROSAMINES

LITERATURE REVIEW

Biodegradation of Nitrosamines in Soil and Wastewater

Insufficient data are available to predict the rate at which NDMA and other nitrosamines would be eliminated due to microbial degradation in the aquatic environment or in the course of biological steps in water treatment. Microbial degradation of NDMA in aqueous and soil systems has not been clearly established. Reports have rather demonstrated a wide range of non-conclusive findings concerning the degradability of nitrosamines and will therefore be described here in more detail.

In contrast to the water phase, photolysis and volatilization rapidly remove NDMA on soil surfaces. Oliver (1979b) reported that 30-80% of an unreported concentration of NDMA volatilized from the soil within the first few hours after application to the soil surface. Once incorporated into subsurface soil, however, far less of the nitrosamine would enter the atmosphere by volatilization. NDMA will be highly mobile, with the potential to migrate into groundwater supplies (Gunnison et al. 2000). Microbial degradation is the most important removal process of NDMA in subsurface soil, even though it turned out to be rather slow.

In a combined study by Gan et al. (2006) and Arienzo et al. (2006) a field study was conducted in the summer months under arid Southern California weather conditions to evaluate the leaching potential of NDMA in turf grass soils during wastewater irrigation. NDMA introduced via surface irrigation of wastewater contaminated with roughly 1000 ng/L did not leach through the upper 100 cm profile of the turf grass soil. This limited NDMA leaching was attributed mainly to rapid NDMA volatilization near the soil surface (Arienzo et al. 2006). However, as part of these experiments, NDMA biodegradation was studied in laboratory incubation experiments with soil samples from different depths. In non-sterile soils, NDMA showed moderate to long persistence, with half-lifes ranging from 13.5 to 36.5 d in the top 0-10 cm layer, and from 26.3 to 79.7 d in the 10-20 and 20-30 cm layers, respectively. Sterilization of the soil samples by autoclaving revealed significantly longer half-lives compared to the non-sterile soils (Gan et al. 2006).

Tate and Alexander (1975) spiked silt loam with 22.5 mg/kg NDMA at 30 °C and observed a lag of approximately 30 days before slow disappearance from soil commenced. 50% loss occurred after about 55 days and 60% after about 70 days incubation. As part of the same study, 40% loss was observed in 2 days in soil amended with 50 mg/kg NDMA and 44% loss was observed in 5 days in soil amended with 250 mg/kg NDMA. These initial losses were followed by either a very small or no loss over the next 3 weeks (Tate and Alexander 1975). Initial, rapid loss of NDMA was attributed to volatilization and slow, gradual loss of NDMA was attributed to biodegradation. However, the authors reported that they were unable to isolate microorganisms from soil and sewage capable of metabolizing nitrosamines (Tate and Alexander 1975; 1976b). In 72-h incubations, NDMA could not be metabolized by axenic cultures of bacteria (p stutzeri, p fragi, B. mycoides, B. cereus, B. subtilis, C. acetobutylicum, C. pasteurianum, A. picolinophilus, A. suboxydans, and Arthrobacter cultures) (Tate and Alexander 1976b). It was further demonstrated that NDMA was rather persistent under both aerobic and anaerobic conditions.

The authors also demonstrated that toxicity to the microflora does not account for the lack of degradation of these compounds (Tate and Alexander 1976b).

In soil experiments, Mallik and Tesfai (1981) incubated NDMA at 4, 25 and 37 °C and found that at all three temperatures about 20-30% of the added NDMA disappeared within the first 20 days of incubation, but little loss was noted thereafter. Even after 30 days of incubation, over 50% of the added NDMA was retained. The rate and quantity of NDMA loss appeared not to be influenced by the incubation temperature, being rather atypical for microbial degradation processes. This was underlined by the finding that the loss of NDMA in non-autoclaved soil was only marginally higher than in autoclaved soil. The influence of soil texture on the rate of disappearance of NDMA was found to be minimal. The authors reported that the rate of loss was also found to be slightly higher in aerobic soil at field capacity compared to super saturated (anaerobic) soil, what might be also caused by a lower volatilization rate in the latter case. After a 30-day incubation period, 60% of added NDMA remained in soil at field capacity and 70% of added NDMA remained in super saturated soil (Mallik and Tesfai 1981).

In comparable experiments with NDPA, this nitrosamine disappeared faster than NDMA (Mallik and Tesfai 1981). The results indicated that the soil with the better aeration favored the disappearance of NDMA. These findings are in agreement with results of Mosier and Torbit (1976) who found that in cattle manure 61% of added NDMA was lost after 28 days of incubation when the sample was kept continuously stirred (increased aeration), whereas under static condition that loss was only 25% during the same period. Furthermore, also Tate and Alexander (1976b) noted very little loss of NDMA from flooded soil. In soil amended with wheat straw, which increased the content of organic matter from 2.2 to 18%, disappearance of the nitrosamines was accelerated substantially, with over 60% of the added NDMA being lost by day 15 (Mallik and Tesfai 1981). It was concluded, that the higher rate of loss of NDMA in amended soil compared to unamended soil was due to higher microbial activity resulting from added organic matter. However, the possibility of adsorption of a part of the nitrosamines by the soil organic matter could not be ruled out (Mallik and Tesfai 1981).

Yang et al. (2005) studied the degradation of NDMA in three different landscape soils through incubation experiments under laboratory controlled conditions. NDMA removal occurred at different rates among these soils and half-lives ranged from 4.1 to 22.5 days at a temperature of 21 °C. For each soil, sterilization by autoclaving and incubating at a lower temperature of 10 °C resulted in a substantially prolonged persistence of NDMA, indicating that NDMA was degraded biologically. Furthermore, the relative persistence was inversely correlated with soil organic matter content, soil microbial mass, and soil dehydrogenase activity, also suggesting the importance of microorganisms in NDMA degradation in these soils. On the other hand, the initial NDMA concentration (250 μg/kg or 25 μg/kg) or addition of other anthropogenic nutrient sources had little or no effect on NDMA persistence.

Yang et al. (2005) also used the most probable number (MPN) method in an attempt to enumerate NDMA-degrading bacteria by using NDMA as the sole substrate. However, no significant NDMA degradation was found in NDMA-amended culture tubes even four weeks after the treatment. The authors concluded that although NDMA was degraded biologically, the degradation was not caused by degraders that used NDMA as their growth substrates. Therefore, microbial degradation of NDMA in these soils was attributed to co-metabolic effects.

Oliver et al. (1979) investigated the degradation of ^{14}C-labeled nitrosamines including NDMA, NDEA, and NDPA in aerobic soils. $^{14}CO_2$ evolution proceeded at the same rate for all nitrosamines investigated. Interestingly, for NDPA $^{14}CO_2$ evolution did not depend on the location of the label in the molecule. No $^{14}CO_2$ evolution was observed in sterile soils. The rate of $^{14}CO_2$ production did not vary with NDPA concentration over the range of 5 µg/kg to 10 mg/kg. The authors determined a half-life for NDMA of 20 days at 23 °C. Losses of nitrosamines were attributed to both volatilization and biodegradation.

Kaplan and Kaplan (1985) demonstrated that NDMA was mineralized by microorganisms in aqueous (lake water) and soil systems. In both systems, the degradation was found to be rather slow. However, rate constants and total percentage of mineralization increased with decreasing initial concentrations of NDMA. In laboratory-scale simulated trickling filters containing an activated charcoal bed with an active biofilm, mineralization of NDMA after 114 days amounted in the lower concentration range (ca. 50 µg/mL) to 24-72%, depending on the nutrient status of the lake water. It was found that rates of mineralization were reduced in aqueous systems when supplemental carbon was available, indicating that supplemental carbon was utilized preferentially to NDMA. In soils, however, percentages of organic matter and supplemental carbon had little effect on rates of mineralization. In studies with different soil moisture levels ranging from dry soil over 25, 50, 75 to 100% of saturation, almost all mineralization activity was inhibited in dry soil, whereas the amount of NDMA mineralized was similar at all other moisture levels. In these experiments, NDMA was also found to be non-toxic to natural populations of microorganisms at concentrations of up to 10 mg/mL (Kaplan and Kaplan 1985).

In a static culture flask biodegradability test based on the biochemical oxygen demand (BOD), Tabak et al. (1981) could find only a slow degradation of NDPA (40-50% after 28 days) under the test conditions established (5 mg yeast/L, 5 mg/L test compound).

Under anaerobic conditions, NDPA degraded slowly in sandy loam with 77% and in silty loam with 50% of the initially applied NDPA remaining after 60 days and 47 days, respectively (Saunders et al. 1979). Under aerobic conditions, NDPA dissipated in these two soils to less than 10 and 1% of the initially applied amount after 69 days. However, much of the dissipation in the aerobic experiment was attributed to volatilization (Saunders et al. 1979).

Duré et al. (1978) investigated the biological degradation of NDMA, NDEA and NDPA in laboratory-scale experiments with soil under saturated anaerobic conditions. It was found that biodegradation rates decreased in the order NDMA > NDEA > NDPA with half-lives between 448 and 2500 days.

According to the studies by Tate and Alexander (1975), biotransformation and biodegradation constitute no significant factors affecting the fate of NDMA in lake water. NDMA, NDEA and NDPA did not disappear from samples of lake water incubated at 30 °C over a period of nearly 4 months. Howard et al. (1991) reported a half-life range for NDMA in groundwater of 42-360 days, based on unacclimated aqueous aerobic biodegradation.

Yordy and Alexander (1980) investigated the microbial degradation of NDELA in samples from sewage and two lake waters. The rate of NDELA disappearance was generally slow. In the samples of lake water, rate and extent of NDELA metabolism varied seasonally, and no disappearance occurred in samples taken in winter. The products formed from NDELA in lake water turned out to be quite persistent in this matrix and no mineralization was observable for up to 55 days. In sewage, no seasonal effect of the microbial conversion was evident, and the products of metabolism were slowly mineralized. However, CO_2 evolution was not observed until about day 25. 40% of the carbon added as NDELA was recovered as CO_2 when the experiment was terminated after 90 days. The authors concluded that NDELA mineralization involves most probably two microbial populations, one converting NDELA to the organic products, and a second oxidizing these intermediates to CO_2.

Mosier and Torbit (1976) investigated the stability of NDMA in manure and water extracts of manure. The nitrosamine was found to be stable in these matrices under both aerobic and anaerobic conditions for 1 month (Mosier and Torbit 1976).

In biodegradation experiments with soil by Gunnison et al. (2000) first order kinetics were established for NDMA degradation. Within a time frame of 30 days, loss of NDMA (initial concentration: 50 µg/L) by microbial degradation (measured as $^{14}CO_2$ evolution) in aerobic and anaerobic incubations with subsurface soils under saturated conditions ranged from about 30% to > 60%. Depending on the soil, there were either identical degradation yields obtained under anaerobic and aerobic conditions or the aerobic mineralization exceeded the anaerobic one. Half-lives of NDMA in various soil slurries ranged from 11.6 to 38.5 days at the 50 µg/L level. Similar to the findings of Kaplan and Kaplan (1985) already mentioned mineralization rates decreased nearly 3-fold when incubations were performed with a higher initial concentration of 50 mg/L NDMA. On the other hand, addition of small to moderate levels of nitrate and phosphate in the course of the incubation caused a significant increase in NDMA mineralization.

Gunnison et al. (2000) also investigated the fate of NDMA in an operating intercept-and-treat groundwater remediation system that extracts NDMA contaminated groundwater from an aquifer, treats it by adsorption onto granular activated carbon (GAC) and subsequently discharges it into a system of groundwater recharge trenches. It was found that the bioactive GAC units were not highly effective in removing NDMA from the system influent, since mean levels of NDMA in the influent and effluent waters were 0.35 and 0.20 µg/L, respectively. Most interestingly, even though NDMA was easily transported through the aquifer up until interception and passed through the GAC treatment system, it could not be detected in the recharged groundwater down-gradient. It was concluded that certain changes of the water matrix that occur within the treatment plant improved NDMA attenuation. Mitch et al. (2003a) proposed that the GAC unit may have removed competitive substrates from solution, improving the microbial degradation of NDMA during artificial recharge.

Several authors reported on the degradability of NDMA in full-scale secondary wastewater treatment or in the laboratory. Pehlivanoglu-Mantas and Sedlak (2006) conducted batch experiments with wastewater effluent and bacteria derived from activated sludge. In these experiments, the initial NDMA concentration between 15 and 130 ng/L decreased by more than 50% within 30 days. No reduction in NDMA was observed in sterile controls, indicating that the losses were attributable to biological processes.

Quesnel and Nakhla (2005) demonstrated that a pilot scale system with conventional activated sludge employed as secondary biological treatment step was able to reduce NDMA from feed concentrations of ranging from 0.04 to 41 µg/L to the method detection limit of 0.04 µg/L.

Sedlak et al. (2005) reported on the incomplete biodegradation of NDMA in full scale municipal wastewater treatment systems with secondary treatment based on activated sludge, with a median influent concentration of 88 ng/L NDMA and a median secondary effluent concentration of 46 ng/L. Depending on the treatment system and the time point of sampling, NDMA removal varied between little or no removal to approximately 93%.

Mallik and Tesfai (1981) tested seven microbial cultures (Escherichia coli, Enterobacter cloacae, Pseudomonas sp., Pseudomonas fluorescens, Serratia marcescens, Staphylococcus aureus and Streptomyces sp.) under axenic culture condition. With the exception of Escherichia coli, these microorganisms could not degrade NDMA at substantial amounts. Also NDMA elimination by Escherichia coli was not complete and amounted to 80%. According to Cattaneo et al. (1981), also Streptobacterium spp, Clostridium perfringens spp, E. coli, and Fusiformis fragilis are not active in degrading NDMA. Three strains of non-pathogenic microorganisms (i.e. Rhizopus oryzae, Streptococcus cremoris, and Sccaromyces rouxii) were found to be active in degrading NDMA, NDEA and NDPA (Harada and Yamada 1979). Thereby, NDPA was degraded more rapidly than the other nitrosamines. In the tests, Rhizopus oryzae degraded 80% of NDPA, 25% of NDEA, and 10% of NDMA. Streptococcus cremoris and Saccharomyces rouxii degraded 70% and established 50% of NDPA, respectively, while NDEA and NDMA were degraded to a smaller extent (10-30%). In cells pre-cultured with NDPA, the degradation of NDPA was slightly improved (ca. 10% higher yield) suggesting that the degradation might rely on an inducible enzyme (Harada and Yamada 1979).

In another study, Harada (1980) demonstrated that NDPA and NDMA degradation in cell-free extracts of Rhizopus oryzae proceeded both aerobically as well as anaerobically, but with higher activities under anaerobic conditions. Yoshinari and Shafer (1990) studied the degradation of ^{14}C-NDMA by the methanotroph Methylosinus trichosporium OB3b and observed that this organism was capable of both assimilating NDMA-carbon and respiring it as $^{14}CO_2$. Rates of $^{14}CO_2$ production from NDMA and its cellular uptake were linearly correlated with NDMA concentrations. Approximately 3% of the added NDMA was metabolized within 24 h. The authors suppose that the metabolism of NDMA was carried out by methane monooxygenase.

The metabolism of NDMA in mammalian cell systems has been well documented, with nitrogen gas, formaldehyde, formic acid, methylamine, N-methylhydrazine, N-methylhydroxylamine, and N,N-dimethylhydrazine identified as intermediates in various in vitro studies (Grilli and Prodi 1975; Lottikar et al. 1975; Lake et al. 1976; Milstein and Guttenplan 1979; Jensen et al. 1981; Godoy et al. 1984; Yoshinari 1995). Three pathways of degradation of NDMA by microbial enzymes and by microsomal P-450 monooxygenases were established: denitrosation, demethylation and a reductive route (Figure 6.1).

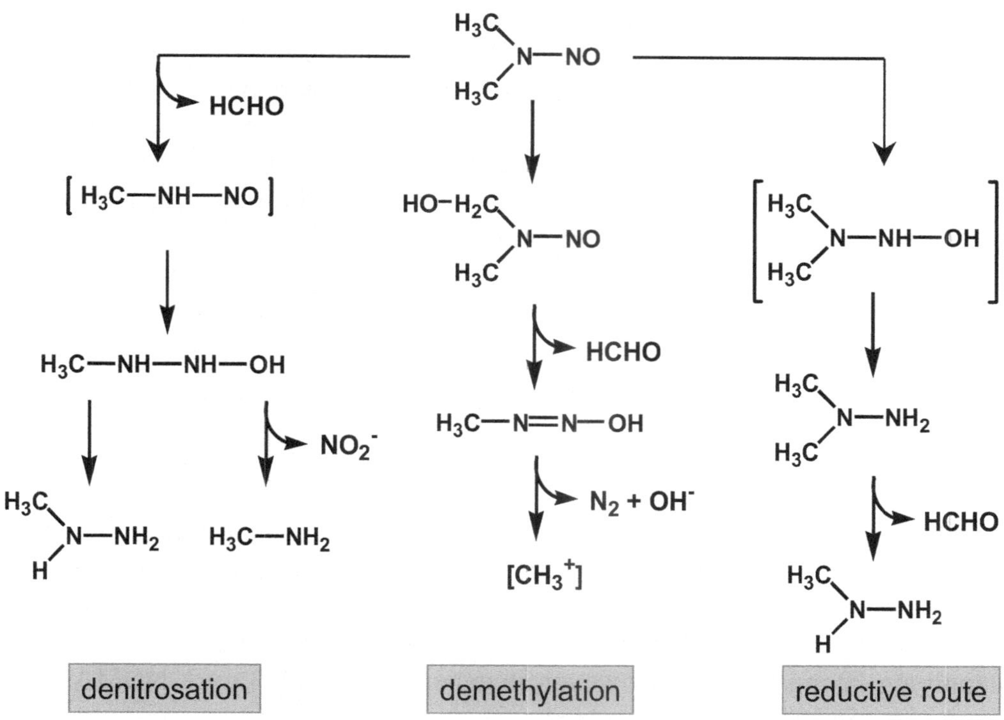

Figure 6.1 Biodegradation pathways for NDMA identified in mammalian systems

Sharp et al. (2005) showed that bacteria expressing monooxygenase enzymes functionally similar to those demonstrated to degrade NDMA in mammalian cells have the capability to degrade NDMA. Particularly, induction of the soluble methane monooxygenase (sMMO) expressed by Methylosinus trichosporium OB3b, the propane monooxygenase (PMO) enzyme of Mycobacterium vaccae JOB-5, and the toluene 4-monooxygenase found in Ralstonia picketti PKO1 and Pseudomonas mendocina KR1 caused NDMA destruction by these strains. In each of these cases, brief exposure to acetylene gas, a suicide substrate for certain monooxygenases, inhibited the degradation of NDMA. Since also monooxygenase enzymes catalyze NDMA degradation in mammalian cells, bacterial NDMA degradation may occur by a similar biochemical degradation pathway (Sharp et al. 2005). Unfortunately, the microbial metabolism of NDMA in aqueous or soil systems has not been established in more detail. Only Kaplan and Kaplan (1985) identified two metabolic intermediates, formaldehyde and methylamine, from the microbial degradation of NDMA, supporting the analogy between bacterial and mammalian biochemical pathways.

NDMA is expected to be highly mobile in soil and it has the potential to leach into the groundwater compartment (Dean-Raymond and Alexander 1976; Greene et al. 1981; ATSDR 1989; Gunnison et al. 2000; World Health Organization 2002; Yang et al. 2005). Indeed, several authors have reported on contamination of groundwater sources by NDMA, in most cases attributed to emissions of nearby chemical facilities (World Health Organization 2002; ATSDR 1989; Xiang et al. 1996). Adsorption of NDMA, NDEA and NDPA onto soils or aquifer solids is rather insignificant (Duré et al. 1978; Saunders et al. 1979; Gunnison et al. 2000; Yang et al. 2005). In experiments with abiotic soil columns, NDMA, NEMA, NDEA, NDPA, NDBA, NPYR and NPIP were not retained by the soil (Hoppe et al. 2006). NDMA was even reported to move as readily as chloride through soil (Dean-Raymond and Alexander 1976).

Most of the presented experiments were generated in laboratory vessels and predictions for the field must be treated with caution. A comprehensive evaluation of all results leads to the assumption that nitrosamines are rather persistent in the aquatic environment. It has been clearly demonstrated that nitrosamines can be mineralized by microbial communities under certain conditions. However, this process is, when it occurs, relatively slow and levels of nutrients and cofactors seem to be of major importance for successful nitrosamine elimination. Reported half-lives for nitrosamine biodegradation in soil and water are in the range from several days over months to even years. It is still unclear, how nitrosamines like NDMA behave during bank filtration, artificial recharge or biofiltration units used for drinking water treatment and it has to be clarified if such treatment steps might be an option for NDMA removal under field conditions.

Ozonation and Other Oxidation Processes

Ozonation

The pure application of ozone shows no effect on NDMA concentration neither under atmospheric nor aqueous conditions (Wong 2001; Sharpless and Linden 2003). Recent work, however, gave evidence of a small decline of NDMA concentration during ozonation in aqueous solutions (Mofidi et al. 2002). A 20% reduction of an initial NDMA level of 3,000 µg/L was achieved when applying 10 mg/L ozone. However, in experiments with 3 µg/L NDMA no elimination was observed when 5 mg/L ozone was added.

Ozone/H_2O_2 Treatment (Peroxone Process)

Often, the effectiveness of ozonation can be improved by addition of H_2O_2, due to an enhanced formation of $^{\bullet}OH$ radicals. In the literature, a value for the pseudo-first order rate constant for the reaction of NDMA with $^{\bullet}OH$ radicals of $3.3 \times 10^8\ M^{-1}s^{-1}$ is given (Wong 2001; Sharpless and Linden 2003). Recently, Mezyk et al. (2004) reported a similar value for the rate constant with $4.3 \times 10^8\ M^{-1}s^{-1}$ based on γ-radiolysis experiments. Mofidi et al. (2002) demonstrated a 90% removal of NDMA when applying 15 mg/L ozone and H_2O_2 in a 1:1 ratio for a solution containing 3 μg/L NDMA. 5 mg/L ozone/H_2O_2 (1:1) caused 50% removal. In groundwater, only a slight decline in removal efficiency was observed under the same conditions. Moreover, in the range between 0.5 and 3 μg/L, the removal efficiency was independent from the initial NDMA concentration.

Oxidation by Fenton-Type Reactions

Extensive studies have been carried out concerning the removal of nitrosamines by the Fenton reaction. The products of Fenton reagent degradation of NDMA were identified as formaldehyde, methylamine (MA), and nitrate or nitrogen gas, depending on the reaction conditions (Wink et al. 1991, 1994). During the Fenton reaction ferrous ion reacts with H_2O_2 to form ferric ion and $^{\bullet}OH$ radicals.

$$Fe^{2+} + H_2O_2 \rightarrow Fe^{3+} + 2\ OH^*$$

Heur et al. (1989) showed a stoichiometric conversion of NDMA to MA by Fenton oxidation. They proposed a denitrosation mechanism based on self-decomposition of the carbon-centered radical which is produced by hydrogen abstraction from the methyl group of NDMA by attack of $^{\bullet}OH$ (Figure 6.2). Recently, Mezyk et al. (2004) supported the production of the carbon-centered radical from NDMA oxidation by $^{\bullet}OH$ using γ-radiolysis experiments and DMPO spin trapping EPR technique.

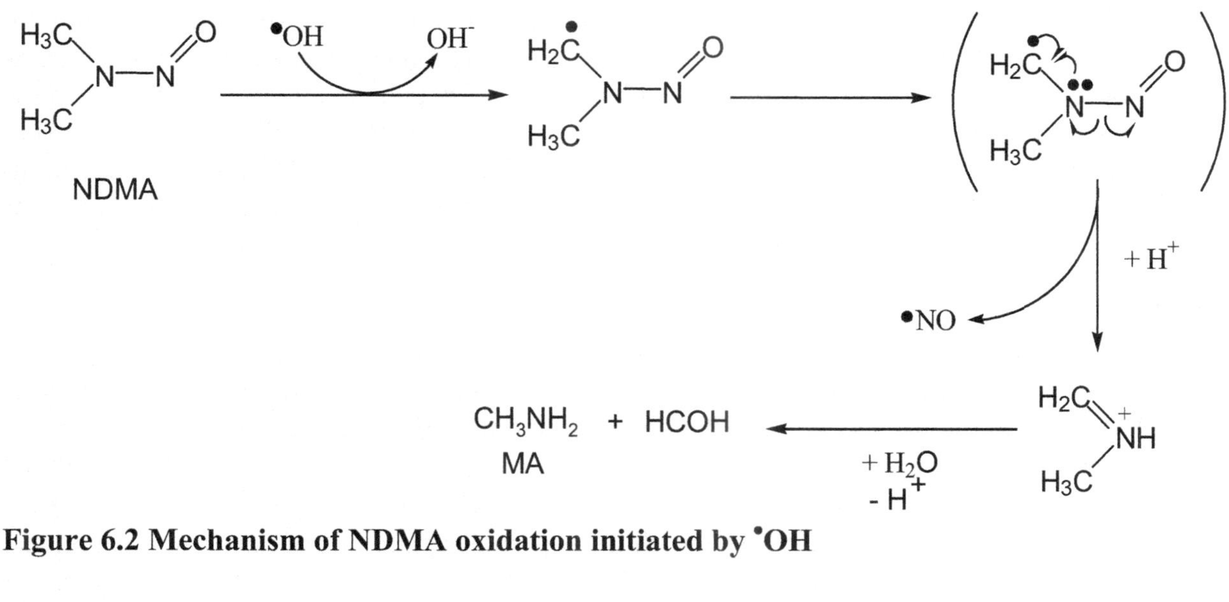

Figure 6.2 Mechanism of NDMA oxidation initiated by $^{\bullet}$OH

It has been postulated that •OH radicals are the exclusive oxidizing intermediate formed during the Fenton-type reaction. However, Wink et al. (1994) showed the existence of an intermediate X (assumed to be a peroxo-ferrous complex) which reacts faster with NDMA than the •OH radicals. X reacts with NDMA by hydrogen abstraction yielding the alkylnitrosoamino radical. This radical decomposes to nitric oxide which is then bound to ferrous ion. The latter can then decompose to form nitrate or, at lower concentrations of NDMA, to form an additional intermediate Y (probably Fe(IV)O) which in turn reacts with NDMA again. The second elimination is suggested to take place via electron transfer instead of hydrogen abstraction.

Mochizuki et al. (1994) studied the product formation of NDMA, NDEA, NMPA (N-nitroso-N-methylpropylamine), NDBA and NMBA (N-nitroso-N-methylbutylamine) during the Fenton-type reaction of N-nitrosodialkylamines in more detail. As reagent, a combination of $Fe^{2+}/Cu^{2+}/H_2O_2$ was used. Experiments were conducted at pH 4.5, 37 °C and with a duration of 2 hours. The investigators observed an increase in mutagenicity during Fenton catalyzed destructtion of NMBA, NDPA, NDBA, NMPA, but not for NDMA and NDEA. Fe^{2+} and Cu^{2+} turned out to act synergistically in these reactions. The mutagenic product of NMBA was characterized as 5,6-dihydro-6 methyl-6-nitrito-4*H*-1,2,3-oxadiazine.

Kommineni et al. (2003) investigated the combined process of adsorption on activated carbon and subsequent regeneration of the carbon by Fenton-type oxidation of NDMA. For investigating the NDMA oxidation via the Fenton reaction, Milli-Q water solutions containing 1 mg/L NDMA were spiked with 2.7 g/L H_2O_2 and different iron concentrations at different pH values. At low pH (< 2.0, no iron precipitation) NDMA was effectively removed within 30 min at an iron concentration of 1.2 g/L. By application of a lower iron concentration of 0.608 g/L about 4 hours were needed for complete NDMA removal, implying that a higher iron concentration leads to a faster degradation of NDMA. At pH 7, a much slower reduction rate was observed, the residual value of NDMA after 24 h still exceeding 700 µg/L. A lower residual level (320 µg/L) was achieved, when EDTA was added to keep the iron in solution. However, EDTA tends to react with •OH radicals itself ($k = 4 \times 10^9\ M^{-1}s^{-1}$) and additionally competes with NDMA for adsorption places on GAC. Finally, adsorption experiments were carried out in a GAC column that was afterwards regenerated by flushing the column with a solution of 2.7 g/L H_2O_2 and 1.8 g/L iron at pH 2 for 24 h. A 99% reduction of adsorbed NDMA was achieved with only little adsorption capacity loss due to the regeneration. A higher amount of H_2O_2 did not lead to an enhancement of NDMA destruction; the accelerated rate of •OH radical formation is diminished by a higher radical scavenging activity of the residual H_2O_2. A rough cost estimation based on the H_2O_2 consumption in these experiments without matrix proved the overall process to be economically feasible. However, one general problem with the application of Fenton-type reactions on natural waters is the radical scavenging activity of natural water constituents, increasing costs due to the necessity of higher input levels of H_2O_2.

Oxidation by UV-A/TiO₂

Lee at al. (2005c) investigated the photocatalytic degradation of NDMA in a UV-A/TiO_2 system, focusing on the influence of the TiO_2 surface modification on NDMA degradation pathways. In addition to the reaction products observed in experiments with •OH radicals, also DMA was formed by attack of •OH to the nitroso group. However, this was possibly a result of the interaction of the oxygen atom of NDMA with TiO_2 particles. The interaction of the negatively charged oxygen atom of the nitroso group of NDMA with protons, concentrated at the interface region around TiO_2 particles, may enable the attack of $^\bullet OH$ to the nitroso group, which is hardly expected in a homogeneous system.

UV Photooxidation

NDMA exhibits two absorption bands with one maximum at 228 nm (molar extinction coefficient $\varepsilon_{228nm} = 7378\ M^{-1}cm^{-1}$) and one at 332 nm ($\varepsilon_{332nm} = 109\ M^{-1}cm^{-1}$) (Stefan and Bolton 2002). The absorption spectrum is illustrated in Figure 6.3. The first absorption maximum is based on a $\pi-\pi*$ intermolecular charge transfer whereas the second, weaker absorption band is caused by an n-$\pi*$ transition band. Both excitations lead, however, to the destruction of the N-N bond.

Polo and Chow (1976) showed that NDMA exhibits a resistance towards photolysis in neutral solutions but is readily removed under acidic conditions (pH 0-4). In a solution with 0.1 N hydrochloric acid they determined a quantum yield of ϕ = 7-8. They postulated the existence of a photo labile protonated species in acidic solutions which is especially susceptible to photo-oxidation (Figure 6.4).

Similar observations were made concerning other N-nitrosamines. In Table 6.1 the molar extinction coefficients for several nitrosamines are given. As can be seen all nitrosamines under investigation exhibit the two adsorption bands at around 230 and 335 nm, a clear indication that the adsorption of UV light is due to the delocated electrons of the N-N bond. At both wave-lengths the molar adsorption coefficients ε of the different nitrosamines range in the same order of magnitude.

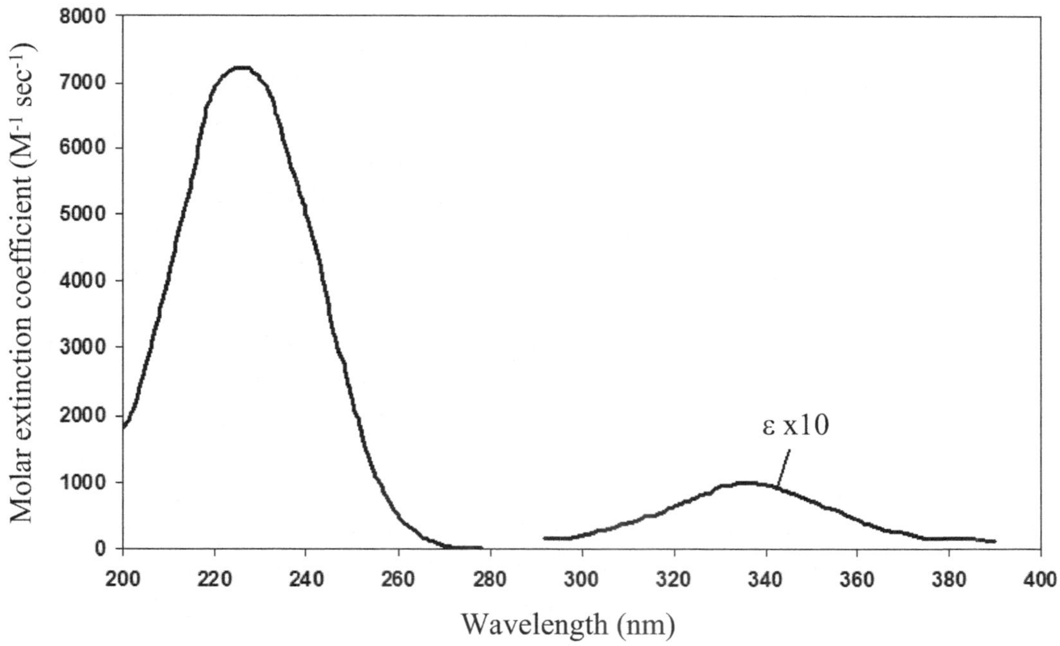

Figure 6.3 Absorption spectrum of NDMA

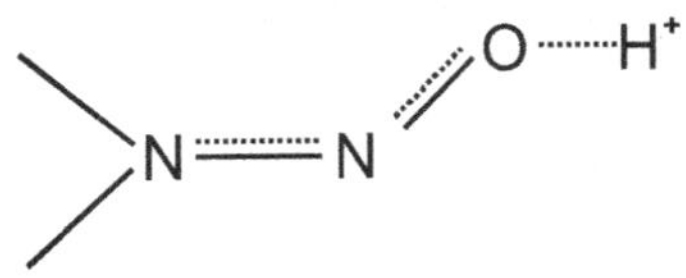

Figure 6.4 Protonated form of NDMA in acidic solutions

Table 6.1
Molar extinction coefficients ε_{mol} at absorption maximum λ_{max}

	λ_{max} (nm)	ε_{mol} (cm^2/mol)	λ_{max} (nm)	ε_{mol} (cm^2/mol)
NDMA	332	109	230	7378
NEMA	333.5	88	227.5	7964
NDEA	338.5	94	241	5326
NDPA	339	99	242	6019
NMBA	334	79	223.5	7488
NMPentA	334	104	230	7645
NPIP	338	99	234.5	10305
NPYR	334.5	116	230.5	9540

Source: Duré et al. 1978

Sunlight Exposure and Atmospheric Destruction

Destruction of NDMA via sunlight exposure mainly occurs by adsorption of UV-A irradiation (wavelength between 300 and 350 nm). Duré et al. (1978) studied the NDMA removal during sunlight exposure in a lab-scale experiment. At 27 °C, 1.6×10^{-3} mol of NDMA were converted in 5 h with an average irradiation flux of 45 mW on the area of 15.2 cm^2. This yields a calculated quantum yield of approximately 0.7. However, the application of these considerations to real surface water still leads to a half-live of approximately 250 days for NDMA.

The technique was applied in pilot scale in a water treatment plant in Orange County in California where two wells were contaminated with NDMA at a level of 100 ng/L. In lab-scale tests raw water spiked at 600 ng/L NDMA was exposed to sunlight for 3 h, after which no NDMA was found. For pilot scale tests open concrete basins were used with a theoretical retention time of 1.5 h. An average of 40% destruction was achieved in pilot scale (Soroushian et al. 2001; Mitch et al. 2003a). Atmospheric gaseous reactions of NDMA are fast and efficient. NDMA is removed within a few hours by UV adsorption, photooxidation and reactions with ozone or $^{\bullet}OH$ radicals in the gas phase. The quantum yield for gaseous reactions of NDMA at wavelengths > 290 nm reaches unity. A major reaction product is dimethylnitramine which is not further degraded by photolysis. The half-life of NDMA for atmospheric reactions with $^{\bullet}OH$ radicals is approximately 3 days in contrast to 5 min for direct photolysis (Tuazon et al. 1984; Mitch et al. 2003a).

The photolysis of NDPA in lake water was investigated by Saunders and Mosier (1980). Experiments with lake water spiked with NDPA and exposed to sunlight showed a rapid degradation, independently from the initial NDPA concentration (1 and 10 mg/L, respectively). Furthermore, no impact was found for the oxygen content of the water in laboratory-scale experiments. In contrast to other studies, the reaction rate of the NDPA degradation was found to be independent from the pH of the water. Identified reaction products were di-n-propylamine (DPA) and n-propylamine (NPA).

Photooxidation Kinetics

The photooxidation of NDMA has been studied in detail in several publications. Early experiments concerning NDMA elimination by UV irradiation with a high pressure mercury lamp by Duré et al. (1978), starting with an initial concentration of 50 mg/L, revealed first order degradation kinetics.

Polo and Chow (1976) determined two-phase reaction kinetics for NDMA photodegradation. Starting with an initial NDMA concentration of 75 mg/L, the reaction followed zero-order kinetics during the first half of the experiment and then changed to first-order kinetics. This was confirmed for high initial concentrations of NDMA (up to 74 mg/L) also by Stefan et al. (2002), but not for lower concentrations (74 µg/L). In the latter case pseudo-first order reaction kinetics were observed for the whole experiment, what was also established by other investigators (Liang et al. 2003; Sharpless and Linden 2003) some of which investigated even lower initial concentrations (down to 3 µg/L). A faster degradation rate was observed at pH 3 than at pH 7. Stefan and Bolton (2002) attributed this to a higher quantum yield at low pH values and possible reformation of NDMA in neutral solutions. However, no specific data on quantum yields were given in that study. Duré et al. (1978) found a reaction rate dependency on pH as displayed in Figure 6.5, showing a sharp decline with a minimum at pH values around 9.5 to 11.0. They investigated five different N-nitrosamines, namely NDMA, NDEA, NDPA, NPYR and NPIP at a concentration level of 50 mg/L and each nitrosamine showed a distinct minimum of the reaction rate constant. k_{Phot} values ranged from 2.6 to 4.1 x 10^{-2} min^{-1} in acidic to neutral solutions and from 0.9 to 1.9 x 10^{-2} min^{-1} at the minimum. Considering naturally occurring pH values, however, the minimum is out of range and reaction rates should be in the higher range in natural waters.

An evaluation of the data led to a first estimation of the quantum yield for the destruction of NDMA (Duré et al. 1978). For experiments with sufficient oxygen supplied, the quantum yield was determined to 0.038.

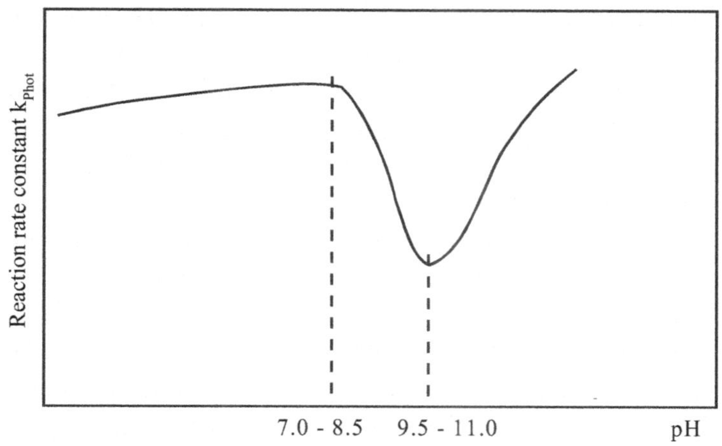

Source: Qualitatively according to Duré et al. (1978)

Figure 6.5 pH dependence of reaction rate constants for N-nitrosamines

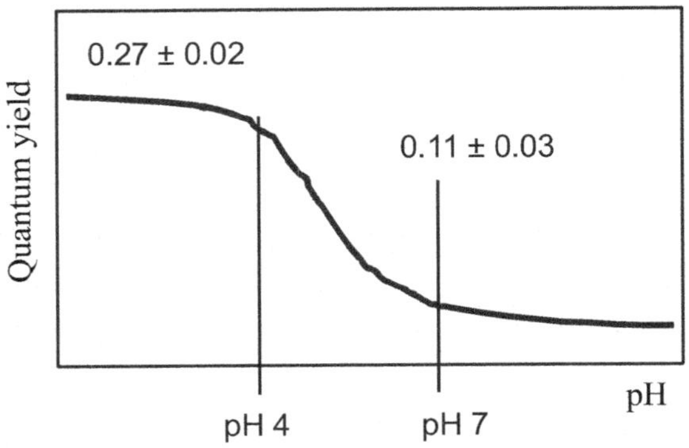

Source: Qualitatively according to Ho et al. (1996)

Figure 6.6 pH dependence of NDMA quantum yield

Ho et al. (1996) determined quantum yields for the photolysis of NDMA with Xenon flash lamps within a pH range from 2 to 14. Initial NDMA concentrations varied between 1.6 and 176 mg/L. The authors found no correlation between the pure quantum yield for NDMA photolysis and the initial concentration but a clear dependence on pH (Figure 6.6). At low pH values a quantum yield of approx. 0.27 was determined. With rising pH, however, a decline in the quantum yield was observed down to 0.11 at pH values above 7. This supports the theory of Polo and Chow (1976) postulating the existence of a photo-labile species at low pH values.

Investigations with a low pressure mercury lamp (almost monochromatic emission at a wavelength of 254 nm) showed at pH 6 and an initial NDMA concentration of approx. 100 µg/L a quantum yield of $\phi_{254nm} = 0.298$, whereas a value of $\phi_{200\text{-}300nm} = 0.309$ was determined when broadband irradiation was applied using a medium pressure mercury lamp (Stefan et al. 2002). This indicates that the quantum yield for NDMA photolysis is, within the first adsorption band, independent from the wavelength. Sharpless and Linden (2003) confirmed these results with an experimentally determined quantum yield of 0.300. For the lower-energy band a quantum yield of 0.314 was calculated in good agreement to the one determined for the short wavelength band (Stefan et al. 2002). Based on these findings, the quantum yield can be regarded as almost constant over the absorption spectrum of NDMA. Stefan et al. (2002) found a decline for the quantum yield at high pH values. This decrease from 0.28-0.30 down to 0.16, however, occurred only at pH values above 11. The discrepancy with the data published by Ho et al. (1996) was attributed to an incorrect calculation of light absorption in the earlier study.

Recently, Lee et al. (2005b) thoroughly investigated the NDMA photolysis under UV irradiation at 253.7 nm, by determining the quantum yield of NDMA photolysis in the pH range of 2–11 under O_2 or N_2 saturation. Figure 6.7 displays the determined quantum yields. Obviously, the pH dependencies of the quantum yields were clearly different under O_2 and N_2 saturation. At pH values between 2 and 8, the quantum yields determined under O_2 saturation were almost constant at around 0.3, while those under N_2 condition showed a maximum at pH 4. A drastic drop of the quantum yield was found at pH $\geq$ 8 in both conditions. In alkaline solutions at pH 9–11, low quantum yields for NDMA photolysis (< 0.1) were observed under O_2 condition, whereas no significant photolysis of NDMA occurred under N_2 condition. The pH-dependent photolytic behavior of NDMA was successfully explained by the interaction of NDMA in the photo-excited state with a proton, and pH-dependent secondary reactions between the primary photolysis products (Lee et al. 2005a, 2005b).

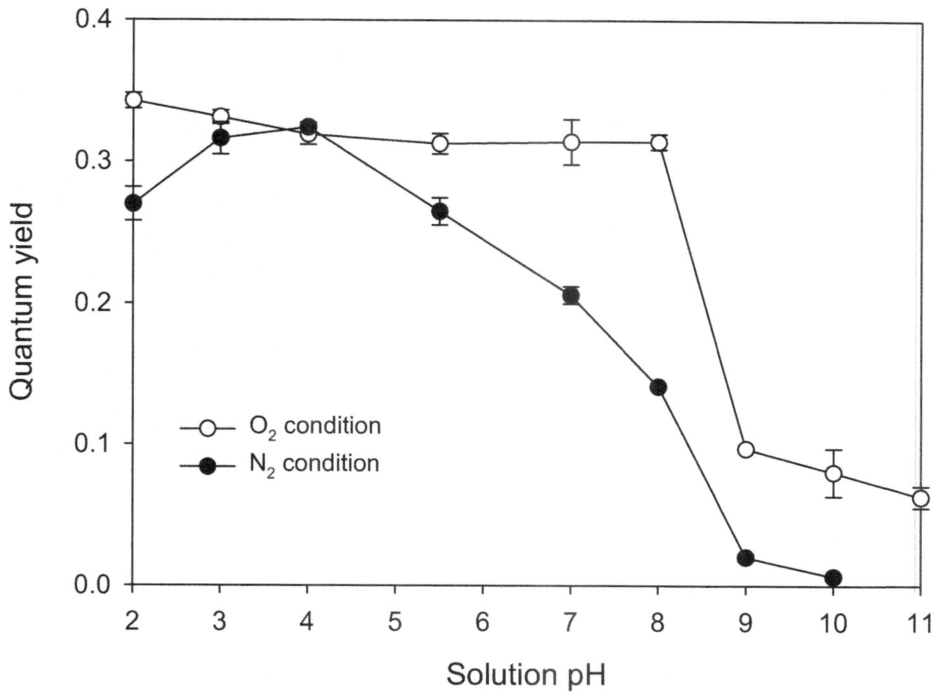

$[NDMA]_0 = 1.0 \times 10^{-3}$ M, $I = 7.1 \times 10^{-6}$ Einstein $l^{-1}s^{-1}$ (253.7 nm)

Source: Lee et al. 2005b

Figure 6.7 Quantum yields for NDMA photolysis as a function of solution pH

The rate constant increases with temperature in an Arrhenius-type manner regardless of pH. Activation energies amount to $E_{A,pH6.5} = 11.5$ kJ/mol at pH 6.5 and $E_{A,pH8.5} = 10.2$ kJ/mol at pH 8.5. A slight decrease in the rate constant is observed as the pH increases. However, considering naturally occurring pH values, one can conclude that there is no significant impact of pH on the photolysis rate of NDMA (Stefan et al. 2002).

Product Formation

Duré et al. (1978) found an almost quantitative conversion of 50 mg/L NDMA and NDEA to their parent amines during UV irradiation with a high pressure mercury lamp. Minor products included nitrate, nitramine, and NO.

Polo and Chow (1976) observed the formation of nitrous acid under acidic conditions at an initial NDMA concentration of 78 mg/L during experiments with two different reactor types. One was equipped with a Pyrex glass cylinder which cuts off low wavelengths and permits NDMA destruction only via the second, weaker absorption band around 340 nm, the other using a quartz glass cylinder with no cut-off of lower wavelengths. The formation rate of nitrous acid was higher during the experiments where the lower wavelengths were present. However, continuing irradiation led to a decrease in nitrous acid formation due to secondary photodecomposition.

Stefan and Bolton (2002) confirmed the formation of nitrite as major reaction product besides dimethylamine by photooxidation of NDMA under acidic conditions (initial nitrosamine concentration 4 mg/L). Nitrate was formed at low levels based on the photooxidation of nitrite to nitrate. Another by-product of NDMA irradiation was formaldehyde, which was oxidized to formic acid during irradiation. After 10 min, 1 mM NDMA was totally degraded and 0.9 mM DMA were formed. The DMA concentration remained constant throughout the further experiment, indicating its photo-stability. Nitrate was initially formed at 0.18 mM, the concentration increasing to 0.3 mM after 100 min of UV irradiation. Correspondingly, nitrite was formed at a maximum concentration of 0.65 mM after total NDMA destruction declining to 0.5 mM after 100 min. Formaldehyde occurred initially at approx. 0.1 mM and was converted to 0.05 mM formic acid. The formation of DMA and nitrite during photo-decomposition of NDMA might lead to NDMA reformation after treatment.

Under neutral conditions the total destruction of NDMA at the same concentration level required an irradiation time of 60 min (Stefan and Bolton 2002). In this experiment, the same major byproducts were identified (DMA, nitrite, and nitrate). Formaldehyde was formed at a slightly higher rate than at pH 3. In contrast to the experiments at pH 3, where no nitrogen loss was observed, the total N-based mass balance revealed a nitrogen loss of about 20% at pH 7. This was explained by the formation of nitrogen-containing gases, such as N_2O, NO and N_2 (Stefan and Bolton 2002). However, it has to be stated that no complete mineralization was achieved under either conditions.

Stefan and Bolton (2002) proposed a pathway for NDMA destruction which is, under weakly acidic conditions, catalyzed by protons. The photolysis of the protonated form of NDMA (compare Figure 6.4) leads to protonated DMA, nitrous acid or HNO and protonated CH_3-N=CH_2. The latter forms methylamine and formaldehyde, with formaldehyde undergoing further oxidation (Stefan and Bolton 2002).

Recently, Lee et al. (2005a and 2005b) reinvestigated the photolytic mechanism of NDMA decomposition during UV irradiation at 253.7 nm. In these investigations, the authors proposed novel photolytic mechanisms for NDMA decomposition induced by $\pi-\pi*$ transition. In the first study (Lee et al. 2005a), two pathways of NDMA photolysis to MA (pathway 1 in Figure 6.8) or DMA (pathway 2 in Figure 6.8) were described, being strongly dependent on the initial NDMA concentration and solution pH, and the nitrite ion produced from the NDMA photolysis was identified as a key reagent in directing the NDMA photolysis toward DMA production. During NDMA photolysis via pathway 2, DMA is produced as a result of an N-N bond fission which is induced by the attack of the nitrite ion on the excited state of NDMA.

In their second study, Lee et al. (2005b) found that the UV photolytic mechanism of NDMA is strongly dependent on solution pH and dissolved oxygen. The direct oxidation of photo-excited NDMA by dissolved oxygen was suggested as a novel photolytic pathway to explain the observed effect of dissolved oxygen (pathway 3 in Figure 6.7). N-Methylformamide (NMF) was observed as a novel photolysis product of NDMA in the neutral pH region under N_2 saturation. The pH-dependent photolytic behaviors of NDMA were successfully explained by interaction of the photo-excited NDMA with a proton, and pH-dependent secondary reactions between the primary photolysis products.

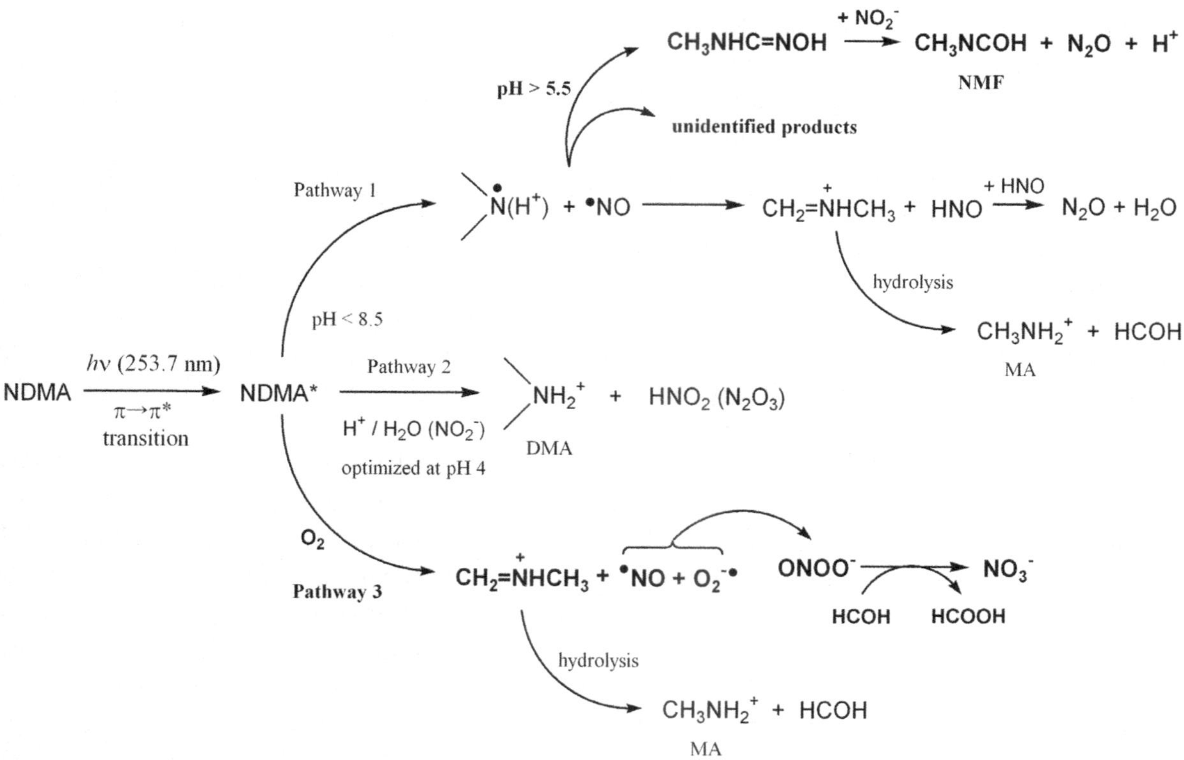

Source: Lee et al. 2005b

Figure 6.8 Overall mechanism of the photolytic pathways and subsequent reactions during NDMA photolysis in aqueous solution

Influence of Water Quality Parameters on the Photolysis of NDMA

The impact of natural water composition, such as NOM content, alkalinity, temperature, and pH on UV photolysis of NDMA was investigated in several studies.

Duré et al. (1978) showed the impact of oxygen on the photolysis of 50 mg/L NDMA. Measured reaction rate constants $k_{Phot.}$ increased from 0.001 min^{-1} in dematerialized oxygen-free water to 0.018 min^{-1} in aqueous solutions containing 12 mg/L oxygen. Further experiments with addition of humic acids, hydroquinone, cumol, and lignin sulfonic acid showed that a radical chain mechanism as reaction mechanism could be ruled out by the fact that an addition of cumol (a radical scavenger) had no influence on the reaction rate ($k_{Phot.}$ = 0.018 min^{-1}). The electronically excited nitrosamine molecule is able to transfer the nitroso group onto hydroquinone and lignin sulfonic acid resulting in an enhancement of the reaction ($k_{Phot.,hydroquinone}$ = 0.029 min^{-1} and $k_{Phot.,lignin\ sulfonic\ acid}$ = 0.021 min^{-1}). The same reaction mechanism takes place with humic acids. However, humic acids absorb light in the 330 nm range. Thus absorption competition leads to a decrease in the net reaction rate ($k_{Phot.,humic\ acid}$ = 0.008 min^{-1}).

Stefan et al. (2002) investigated the impact of NOM and alkalinity on NDMA photodegradation starting with an initial NDMA concentration of 100 µg/L. Due to an inner filter effect of NOM in the 200-300 nm range, the rate constant for NDMA destruction decreases as the concentration of NOM increases from 2 ppm to 12 ppm. Here, competitive UV absorption is responsible for the decline in efficiency. The apparent quantum yield, however, increases at higher NOM concentrations. This is attributed to a photosensitized degradation of NDMA induced by NOM photochemistry. At low NOM levels, alkalinity (40 and 240 ppm) has no significant impact on both the rate constant and the quantum yield. However, at higher NOM levels, a higher alkalinity leads to a decrease in quantum yield. This was explained by the scavenging effect of bicarbonate on $^{\bullet}OH$ radicals generated by NOM photochemistry which are responsible for a part of the NDMA destruction (Stefan et al. 2002).

Similar conclusions were drawn by Mofidi et al. (2002) who investigated the effect of NOM on the photolysis of NDMA at concentrations between 0.1 and 3.0 µg/L (Mofidi et al. 2002). Competitive absorption of UV light by the NOM and scattering of UV light due to turbidity are responsible for a decline in NDMA removal efficiency. This was further underlined by experiments by Liang et al. (2003) who compared the performance of UV photolysis of NDMA for different water types (deionized water, groundwater, and river water exhibiting low TOC / low alkalinity, low TOC / low turbidity / high alkalinity / high nitrate content, and high TOC / high alkalinity / high turbidity, respectively).

A comparison of the absorption spectra of nitrite and nitrate with the one of NDMA shows an overlap at low wavelengths. This means that nitrite or nitrate compete with NDMA for the radiation at these wavelengths. Stefan et al. (2002) demonstrated this effect in a study with a medium pressure mercury lamp where higher initial concentrations of NDMA led to a decrease of the NDMA photolysis rate and thus to a decline of the figure-of-merit EE/O (describing the electrical energy required to bring about degradation of one order of magnitude in 1 m^3 of contaminated water (Bolton et al. 2001)). This was attributed to the formation of by-products such as nitrite or nitrate, being in turn dependent on the initial NDMA concentration. Mofidi et al. (2002) also investigated the inhibiting effect of nitrite and nitrate due to light absorption and oxidation of nitrite to nitrate. Similar effects were found by Liang et al. (2003).

UV/H_2O_2 Treatment

The treatment of NDMA with UV irradiation in combination with H_2O_2 addition aims at enhancing NDMA destruction via formation of highly reactive $^{\bullet}OH$ radicals. The rate constant for the reaction of NDMA with $^{\bullet}OH$ radicals was reported with $3.3 \times 10^8\ M^{-1}\ sec^{-1}$ (Wong 2001; Sharpless and Linden 2003).

Ho et al. (1996) determined quantum yields for the combined UV/H_2O_2 treatment of NDMA in aqueous solutions. They found that the quantum yield for NDMA destruction is dependent on the ratio of NDMA and H_2O_2 concentrations. In cases where the ratio $[H_2O_2]/[NDMA]$ is below 10, H_2O_2 does not significantly disturb the photolysis of NDMA and the quantum yield amounts to 0.12 ± 0.03 and is thus close to the value measured during pure photolysis of NDMA. This indicates that most of the UV radiation is absorbed by NDMA and the main reaction pathway is via photolysis of NDMA. However, if the $[H_2O_2]/[NDMA]$ ratio increases above 100, the quantum yield rises to 0.32 ± 0.04. This indicates that most of the UV is now absorbed by H_2O_2 forming $^{\bullet}OH$ radicals and that the direct photolysis of NDMA is of minor importance as reaction pathway. Another hint for the absorbance of the UV radiation mainly by H_2O_2 is the pH independence of the quantum yield between pH 2 and 8. Since the quantum yield of OH radical production from H_2O_2 photolysis is about 0.90, the determined quantum yield for NDMA destruction of 0.32 indicates that only about 40% of the formed $^{\bullet}OH$ radicals undergo reactions with NDMA. These results were confirmed by Sharpless and Linden (2003). Furthermore, these investigators found out that the performance of medium pressure mercury lamps is more affected than that of low pressure lamps. Due to the enhanced irradiation output of medium pressure lamps at wavelengths below 250 nm, the competitive effect of UV absorption is more pronounced for this lamp type (Sharpless and Linden 2003).

Similar results were achieved with polychromatic pulsed UV lamps. The addition of H_2O_2 shows no improvement in NDMA removal at low initial H_2O_2 doses (0.1 to 1 mg/L). This leads to the conclusion that NDMA is primarily removed by photolysis and that the additional production of •OH radicals is not beneficial. At high H_2O_2 doses (79 mg/L), an inhibiting effect on NDMA photolysis was noticed due to the strong competition for UV absorption. Furthermore, the residual levels of H_2O_2 remained high, with only 20% of the initial H_2O_2 amount being removed (Mofidi et al. 2002; Liang et al. 2003).

Variations of temperature (3 °C, 13 °C, and 25 °C) and pH (6.5, 7.5, and 8.5) during UV/H_2O_2 treatment of a model water (3 mg/L NOM and 140 mg/L $CaCO_3$ alkalinity) showed only little effect on the rate constants for NDMA destruction (Stefan et al. 2002). A slight increase was observed at pH 6.5 where at the same alkalinity less bicarbonate ions are available for scavenging •OH radicals generated by UV induced degradation of H_2O_2. The high molar extinction coefficient and the quantum yield of NDMA as well as the competition of the organic matter for •OH radicals and UV light lead to the assumption that high concentrations of H_2O_2 would be needed to overcome these competing effects. Considering the applied doses of H_2O_2 in water treatment processes only little impact of the OH radical driven process is expected.

Though H_2O_2 has either no or even a slightly inhibiting effect on NDMA photolysis, combined UV/H_2O_2 treatment holds one advantage: the presence of residual H_2O_2 leads to less NDMA reformation during subsequent chlorination as residual H_2O_2 leads to dechlorination.

Reformation of NDMA after Treatment

Several studies demonstrated the possibility for NDMA reformation after UV treatment (Soroushian et al. 2001; Mofidi et al. 2002; Liang et al. 2003). A proposed pathway for this reformation is the reaction of dimethylamine with nitrite, both substances being direct products of NDMA photooxidation (Liang et al. 2003).

Polo and Chow (1976) showed that the formation of HNO_2 during UV irradiation of aqueous NDMA solutions leads, when the parent amine is present, to a reformation of NDMA in the dark. This can cause problems since ubiquitous amines (e.g. in sewage) can be nitrosated by HNO_2. One method for scavenging HNO_2 is the addition of urea ($(NH_2)_2CO$) under acidic conditions which leads to the formation of N_2. The addition of hydrazoic acid (HN_3) allows for a very clean and efficient photolysis of NDMA, with no formation of HNO_2, but poses one problem: HN_3 is photo-labile and thus also subject to photolysis.

Typically, NDMA reformation is enhanced when UV-treated waters are chlorinated. After chlorinating UV treated groundwater with 1 mg/L free chlorine and storage at 4-8 °C for 5 days NDMA concentrations roughly doubled (Modifi et al. 2002; Liang et al. 2003). The addition of H_2O_2 caused less NDMA reformation. In deionized water spiked with 3,000 ng/L NDMA and 40 mg/L nitrate prior to UV irradiation, NDMA concentrations of the UV-treated water increased after chlorination with 1 mg/L free chlorine from 58 ng/L to 99 ng/L, i.e. there was still a significant decrease in NDMA concentration although the concentration of NDM were slightly higher than without chlorine. In these experiments with deionized water H_2O_2 had the same inhibiting effect as in the experiments with groundwater (Modifi et al. 2002; Liang et al. 2003).

Soroushian et al. (2001) investigated the reformation of NDMA after treatment with different UV reactor systems in more detail. They irradiated pretreated natural water (pre-chlorinated reverse osmosis permeate) with NDMA concentrations in the range of 170-380 ng/L in a manner that the NDMA concentrations in the UV-treated water used for the reformation experiments were in the range of 1-3.1 ng/L. Samples stored in the dark directly after irradiation showed no concentration change after three days (concentration levels were always below 2.2 ng/L). However, if the samples were chlorinated with 5 mg/L of free chlorine, a 3 to 15fold increase in NDMA concentration was observed immediately after addition of the disinfectant, remaining then almost constant over three days of storage. Chlorination with 5 mg/L chloramine also caused NDMA formation, but the concentration increase was smaller (only 2fold) and proceeded more slowly. This shows clearly that NDMA reformation during chlorination with free chlorine is more prominent than during chloramination. The addition of H_2O_2 prior to irradiation strongly inhibited the reformation of NDMA (chlorination with free chlorine led to a 2-3fold increase (max. 7 ng/L), while chlorination with chloramine led to a maximum concentration of 2 ng/L).

Comparison of Different UV Systems

Table 6.2 gives a compilation of information concerning the efficiencies of different lamp types for NDMA removal by UV irradiation. The EE/O value, i.e. the energy (given in kWh) needed for removal of NDMA of one order of magnitude in one cubic meter, is strongly dependent on the reactor design. For all lamp types, values determined in lab-scale experiments underestimated the removal efficiency of pilot or full-scale devices. The pH dependency of the quantum yield becomes apparent when comparing the EE/O values for medium pressure lamps in lab-scale experiments with demineralized water: At low pH values, the EE/O value is obviously lower than at pH values around 7 (Stefan and Bolton 2002). The operation of low pressure mercury lamps is economically advantageous, since their EE/O values are lower under the same experimental conditions. In terms of total costs, taking into account also lamp replacement and related operating costs (energy for pumps etc.), however, Lem and Eng (2000b) stated that medium pressure UV systems are often more cost effective than low pressure systems.

Based on the information from lab-scale tests, newly developed pulsed mercury lamps show a moderate EE/O for groundwater that is comparable to medium and low pressure lamps. The impact of water quality was studied in more detail by Liang et al. (2003), investigating different water types. The EE/O value for river water containing more organic background material was twice as high as for groundwater (Table 6.3). While addition of H_2O_2 to deionized water at low doses showed almost no enhancing effect on NDMA removal during irradiation with pulsed UV light, addition of high H_2O_2 doses (79 mg/L) caused a doubling of the respective EE/O value. This reflects the earlier made observation that H_2O_2 is capable of inhibiting NDMA photooxidation. Similarly, the presence of nitrate results in a higher EE/O which can not be decisively lowered by H_2O_2 addition. Natural waters typically show elevated EE/O values when compared to deionized waters. In groundwater, for instance, nitrate and other natural water ingredients compete for UV absorption causing a higher EE/O value. The impact is even more pronounced when investigating river water. In both water types, the addition of H_2O_2 in small doses results in a slightly lower EE/O value, implying that H_2O_2 addition in small amounts is not only useful in inhibiting NDMA reformation (as discussed above) but may also reduce operating costs.

Table 6.2
Comparison of different lamp types used in lab-scale and pilot-scale installations

Lamp type	Test scale	pH / Water type	c_0 NDMA (μg/L)	EE/O (kWh/order/m^3)	Reference
LPML	lab	groundwater*	90	2.60	(Jobb et al. 1994)
LPML	pilot	groundwater*	90	1.26	(Jobb et al. 1994)
LPML	pilot	groundwater*	0.60	0.016 – 0.026	(Lem and Eng 2000b)
Wedeco-Ideal Horizon 300 W LPML	pilot	treated wastewater*	~1.0	0.211 0.185 with 5 mg/L H_2O_2	(Soroushian et al. 2001)
MPML	lab	3.0 / n.s.	74,000	1.04	(Stefan and Bolton 2002)
MPML	lab	3.0 / n.s.	74	0.34	(Stefan and Bolton 2002)
MPML	lab	7.0 / n.s.	74,000	4.3 – 29.6#	(Stefan and Bolton 2002)
MPML	lab	7.0 / n.s.	74	0.53	(Stefan and Bolton 2002)
MPML	lab	7.1 / groundwater	0.71	0.29	(Lem and Eng 2000b)
MPML	lab	natural water*	1.0	0.206	(Bolton et al. 2002)
MPML	pilot	groundwater*	n.s.#	0.040 – 0.264	(Wong 2001)
MPML	pilot	groundwater*	53	0.39	(Bircher et al. 1999)
Aquionics 380 W MPML	pilot	treated wastewater*	~1.0	0.303 0.290 with 5 mg/L H_2O_2	(Soroushian et al. 2001)
Rayox® Tower MPML	full scale	groundwater*	0.90	0.042 – 0.055	(Lem and Eng 2000a)
pulsed UV MPML	lab	7.8 / groundwater	2.80 / 0.10	0.77	(Liang et al. 2003)
pulsed UV MPML	lab	8.3 / river water	2.80	1.82	(Liang et al. 2003)

n.s. – not specified, LPML – low pressure mercury lamp, MPML – medium pressure mercury lamp

*) pH value not given in publication

#) depending on reaction kinetics (highest value for longer irradiation time where already most of the NDMA is destroyed and competition with by-products will occur)

Table 6.3
Impact of water matrix on the electrical energy per order (EE/O) of pulsed UV irradiation in laboratory-scale tests

Water type	$c_0(H_2O_2)$ (mg/L)	$c_0(NO_3^-)$ (mg/L)	EE/O (kWh/order/m^3)
Deionized water	--	--	0.48
Deionized water	0.1	--	0.42
Deionized water	1	--	0.37
Deionized water	79	--	1.06
Deionized water	--	40	0.95
Deionized water	1	40	0.87
Groundwater	--	2.0	0.77
Groundwater	0.1	2.0	0.71
River water	--	1.1	1.82
River water	0.1	1.1	1.61

Conditions: c_0(NDMA) = 2,800 ng/L
Source: Liang et al. (2003)

Field Applications

UV photooxidation proved to be an efficient method for eliminating NDMA from polluted groundwater. As a consequence, different systems for groundwater remediation were developed and implemented in the USA and Canada. Table 6.4 gives an overview of plants currently in operation.

The UV dosage required for NDMA removal by one order of magnitude (1-log removal, requires approx. 1,000 mJ/cm^2) is, however, 10 times higher than the dosage required for equivalent virus removal. In other words, the efficient removal of NDMA is always associated with a total disinfection of the treated water. This has to be borne in mind if the application of UV photolysis is considered for drinking water treatment (Bolton et al. 2002; Mitch et al. 2003a).

Oshweken (Six Nations) Water Supply, Canada. In 1986, in treated water from the Oshweken water supply was contaminated with NDMA at levels between 5 and 115 ng/L leading to its ban for consumption. Measurements in the raw water showed almost no positive results indicating that NDMA was generated during drinking water treatment. Therefore, a study was undertaken investigating different water treatment steps and their impact on NDMA elimination or formation.

UV irradiation with low pressure mercury lamps turned out to be quite successful in NDMA removal. In lab-scale experiments a reduction from 89 ng/L NDMA to levels below 5 ng/L was achieved at a UV dosage of 2.6 kWh/m^3. Later, this value could be reduced to 1.3 kWh/m^3 in the full-scale system (Jobb et al. 1994; Mofidi et al. 2002).

Table 6.4
Overview of industrial applications of UV systems for NDMA removal

Installation	System type	Addition of H_2O_2	Flow rate (m^3/min)	$c_{infl.}$(NDMA) (ng/L)	$c_{effl.}$ NDMA [ng/L]
Suburban water (Covina CA)	1 x 15 lamps	no	11.4	90	< 2
LaPuente Valley Water (LaPuente, CA)	2 x 12 lamps	yes, 3 mg/L	9.5	900	< 2
GenCorp Aerojet (Sacramento, CA)	2 x 12 lamps	yes	15.1	160	< 10
NASA (Whitesands, NM)	1x 12 lamps	no	4.2	2,000	< 10
Town of Elmira (Elmira, Canada)	270 kW Rayox®	no	1.6	2,600	< 140
Orange County Water (Orange County, CA)	2 x 15 lamps	no	27.6	100-300	< 2
California Domestic Water (El Monte, CA)	1 x 18 lamps	stand-by H_2O_2 addition	18.9	100	< 2

Source: Lem and Eng (2000a)
$c_{infl.}$ – influent concentration, $c_{effl.}$ – effluent concentration

LaPuente Valley County Water District, California. In 1997, in Baldwin Park, California, three drinking water production wells had to be closed due to a major contamination of the water with NDMA and perchlorate. In order to remove NDMA by UV-light, the Calgon Carbon Corporation was commissioned to set up a Rayox®-Tower equipped with medium pressure lamps. The Calgon Carbon Corporation has been effectively implementing the UV irradiance technology for NDMA removal since 1991.

In 1999, two Rayox® Towers equipped with 12 lamps each were installed and taken into operation. Each tower was characterized by an EE/O value between 0.042 and 0.055 kWh/order/m^3. At a flow rate of 8.71 m^3/min, 8 lamps with an average power consumption of 20 kVA were sufficient to reduce NDMA concentrations from 900 ng/L down to levels of below 2 ng/L in the treated water. The estimated maximum treatment capacity for the Rayox® Tower was reported to be an initial NDMA concentration of 3 μg/L. Within the final operation scheme, H_2O_2 was added at 3 mg/L for enhanced NDMA and concurrent 1,4-dioxane removal (Lem and Eng 2000a, 2000b; Calgon Carbon Corporation 2000).

NASA Johnson Space Center White Sands Test Facility Las Cruces, New Mexico. During the Apollo space program in the 1960s the facility in Las Cruces was intensively used for rocket engine and materials testing. Due to leakage and spilling of rocket fuel during that time, NDMA was formed by the oxidation of liquid rocket fuel constituents in the groundwater. In 2002, a Rayox® Tower system by Calgon Carbon Corporation was installed for treatment of 4.16 m^3/min contaminated groundwater with twelve 30 kW lamps provided by Lynx Ltd. Treatment goal was the reduction of the initial concentration of 2,000 ng/L NDMA down to levels below 10 ng/L. No further information is reported on the performance of that plant (Lem and Eng 2000a; Dominguez 2001).

Orange County Water District (OCWD). Two raw water wells of the Orange County Water District were found to be contaminated with NDMA at concentrations above the interim action limit of 20 ng/L, set by the Californian Department of Health Services. The secondary effluent of the Orange County Sanitation District (influent of Water Factory 21 (WF21)) showed on an average even concentrations of 170 ng/L NDMA. After treatment in WF21 the water is blended with deep well water and then pumped to recharge wells. However, even this blending did not result in a concentration decline in the water below the interim action level. As a consequence, detailed investigations in the removal efficiency of different treatment steps used in WF21 were initiated (Soroushian et al. 2001).

The treatment processes established included lime treatment, followed by chlorination and recarbonation steps. The treated water was then gravity filtrated and the effluent divided into two streams which were either GAC filtrated and afterwards chlorinated or chlorinated and treated by reverse osmosis. Lime treatment resulted in a slight reduction of the initial NDMA concentration (170 ng/L were on average lowered to 158 ng/L), subsequent chlorination and recarbonation, however, increased the average NDMA level to 289 and 314 ng/L (the latter determined after gravity filtration). GAC treatment of the first partial stream only slightly reduced the NDMA concentration (ca. 279 ng/L), the second chlorination effluent exhibiting values of 353 ng/L (on average). Chlorination (323 ng/L) and subsequent reverse osmosis treatment of the second partial stream resulted in a residual NDMA concentration of 282 ng/L. This study clearly demonstrated that NDMA can be formed during water treatment and that conventional water treatment technologies such as GAC are not capable of reducing this contamination.

For an improved elimination of NDMA several treatment options were tested: natural sun-light exposure, oxygen purge, direct UV photolysis, UV/H_2O_2 treatment, and reverse osmosis. While natural sunlight exposure in pilot scale was partially successful, resulting in a 40% reduction of NDMA levels, purging with oxygen gas had no impact on NDMA concentration. For UV photolysis three different UV reactors were investigated: Two low pressure mercury lamps with 250 and 300 W provided by Trojan Technologies and Wedeco-Ideal Horizon as well as one medium pressure mercury lamp with 380 W from Aquionics. The EE/O values for the tested lamps are given in Table 6.4. The addition of 5 mg/L H_2O_2 resulted in a slightly improved NDMA destruction rate. For the reactor provided by Trojan Technologies, EE/O values were only determined for pre-chlorinated water (see next paragraph). Further tests were carried out with the pre-chlorinated reverse osmosis permeate. The total chlorine residual ranged between 2.3 and 8.8 mg/L and measured EE/O values were twice as high as in the tests with the non-chlorinated water. That means, pre-chlorination increases the required UV dose. The reactor provided by Trojan Technologies showed a higher EE/O than the Wedeco-Ideal Horizon reactor. Different reverse osmosis membranes were tested and provided a removal efficiency for NDMA of 12% for cellulose acetate membranes and of up to 40 % for thin film composite membranes (Soroushian et al. 2001).

UV Systems by Calgon Carbon Corporation (Rayox® Tower). Each Rayox® Tower is equip-ped with 20 kW high efficiency medium pressure mercury lamps. In house experiments prior to pilot-scale tests determined the laboratory-scale EE/O which was scaled up with a model calculation for the sizing of pilot and industrial plants. Because of the longer path lengths, the EE/O in the Rayox® Towers was generally lower (less than a third) than the one measured under laboratory-scale conditions.

A comparison of installation and operating costs for medium pressure and low pressure mercury lamps showed similar annual operating costs (including cleaning and replacement of lamps, additional pumps, etc.), however, lower installation and capital costs for medium pressure mercury lamps make them economically more efficient in the eyes of the manufacturer (Lem and Eng 2000a).

Adsorption

The removal of NDMA and other N-nitrosamine compounds via adsorption onto different adsorbent materials was subject of several studies (Duré et al. 1978; Kaplan and Kaplan 1985; Fleming et al. 1996; Gunnison et al. 2000; Zhu et al. 2001; Kommineni et al. 2003).

Due to its hydrophilic characteristics and its low K_{OW} value (-0.57), NDMA only poorly adsorbs onto soil, activated carbon or other sorbents (Gunnison et al. 2000; Mitch et al. 2003a). In soil columns, no retardation was observed. Adsorption distribution coefficients determined by soil adsorption experiments were low compared to other organic substances (0.40 to 1.14 compared to 100,000 for highly adsorbable compounds and 10...100,000 for moderately adsorbable compounds). Desorption experiments showed similar distribution coefficients indicating that NDMA readily dissolves into the aqueous phase as soon as the concentration level is lower than the equilibrium value (Gunnison et al. 2000). In 1978, Duré et al. (1978) determined rate constants for the adsorption of different nitrosamines onto soil from hydrolysis rate constants and the comparison of biologically inactivated systems with activated ones. The adsorption of nitrosamines, however, is so slow that an accumulation in sediments can be excluded and a fast distribution in the aqueous phase will occur.

Ambersorb 572, a carbonaceous resin was determined as best suited for NDMA removal, followed by coconut shell carbon (Mitch et al. 2003a). However, with values for the Freundlich parameters for Ambersorb 572 amounting to K = 9.65 µg/g, 1/n = 1.17 (Table 6.5), the adsorption performance is rather bad and implies high costs due to short regeneration cycles of the resin (Mitch et al. 2003a). Organic compounds commonly treated by carbon adsorption exhibit K values of > 100 µg/g and 1/n values normally are below 0.6. For species with similar values of K, lower 1/n values (flatter isotherms) will maintain greater sorption capacity as the equilibrium liquid-phase concentration decreases (Kommineni et al. 2003). Kaplan and Kaplan (1985) found an adsorption capacity for NDMA on granular activated coconut charcoal (6/14 mesh) of 110 µg/g.

Kommineni et al. (2003) investigated several activated carbon types and determined Freundlich parameters in pure MilliQ water and groundwater with an initial NDMA concentration of 1 mg/L. Rapid small-scale column test (RSSCT) experiments were carried out at an influent NDMA concentration of 1 mg/L and operation parameters were comparable to full-scale GAC packed towers (empty bed contact time (EBCT) = 1.6 min, specific discharge = 221.7 $Lmin^{-1}m^{-2}$). Determined Freundlich parameters for the different carbon types varied in a wide range (Table 6.5). Best performance in experiments with pure water showed the BCP/MOD1 type (K = 63.5 µg/g, 1/n = 0.336), the others exhibited K values between 1.0 and 14.5 µg/g showing only poor capacities for NDMA adsorption and 1/n values around 1 indicating weak binding energies for NDMA. Breakthrough curves with BCP/MOD1 determined by RSSCT resulted in similar adsorption capacities as determined in the isotherm experiments (806 µg/g compared to 667 µg/g). In groundwater, the performance of all tested activated carbons types deteriorated due to competitive action of other water constituents. Of the carbons tested, BCP/MOD1 was the most affected by the groundwater matrix (Table 6.5).

Fleming et al. (1996) investigated four different adsorbents with regard to their ability to remove NDMA (Table 6.1). For the experiments, groundwater from the Rocky Mountain Arsenal was spiked with 5 µg/L NDMA. With the exception of Ambersorb 572, determined adsorption capacities were rather low (K < 10 µg/g) for all coal types (including the F400 type). Ambersorb 572 was identified as the adsorbent with the best adsorption capacity (K_{max} = 28.37 µg/g), that is, however, based on the determined 1/n value of 2.25, significantly affected by the NDMA concentration in the influent.

Zhu et al. (2001) tested zeolites with regard to their ability to remove NDMA, NPYR and NHMI (N-nitrosohexamethylene imine) from methylene chloride or aqueous solution at high concentration. The following zeolites were studied: NaY, H-ZSM-5, Na-ZSM-5 and NaA. Initial concentrations of nitrosamines for the isotherm determination according to Freundlich varied from 500 to 7500 mg/L. In a first step, adsorption capacities were determined for the methylene chloride solutions. NaY was identified as having the largest adsorption capacity of the three zeolites under investigation (Freundlich parameters: K = 0.710 µg/g, 1/n = 0.738). Differences between NaY and the other zeolites became more pronounced in the order NDMA > NPYR > NHMI. These results can at least partially be explained by different specific surface areas, surface acid-basicities and pore sizes of the zeolites, the latter playing the most important role for the adsorption performance. NaY is the zeolite with the largest pore size and surface area of the zeolites under investigation and thus exhibits the best absorption performance. H-ZSM-5 has the same specific surface area and pore size as Na-ZSM-5 but shows a better adsorption tendency for nitrosamines due to its higher acidity. Comparison with data for adsorption on silica and alumina showed much better absorption for NDMA and NPYR on NaY due to the higher surface area of the zeolite, a special pore structure and the existence of geometric and energetic fields (Zhu et al. 2001). When turning to aqueous solutions, adsorption of nitrosamines is influenced by the water molecules. Since NaY possesses a hydrophilic nature, water molecules are likely to adsorb onto the zeolite blocking sites and channels for NDMA adsorption. Furthermore, additional solute-solvent interactions occur between NDMA and water molecules making it more difficult for NDMA to adsorb onto NaY. As a consequence, more NDMA can be adsorbed on NaY when the adsorbate is dissolved in methylene chloride than in water. In contrast to the results gained from experiments with methylene chloride as solvent, in aqueous solutions Na-ZSM-5 performs best be-cause of its higher hydrophobicity (Zhu et al. 2001). Once adsorbed, the adsorption of nitrosamines to zeolites was found to be quite strong. During thermal regeneration of the zeolites almost no desorption was noticed (only around 2%) and most of the adsorbed nitrosamine molecules were decomposed on the zeolite above 280 °C (Zhu et al. 2001).

Table 6.5
Tested adsorbents and Freundlich parameters for NDMA

Adsorbent	Description	Freundlich parameters K [μg/g] MilliQ	K [μg/g] GW	1/n MilliQ	1/n GW	Reference
Filtrasorb F400	industrial grade carbon	14.5	n.d.	0.602	n.d.	(Kommineni et al. 2003)
Filtrasorb F300	industrial grade carbon	1.00	n.d.	1.138	n.d.	(Kommineni et al. 2003)
BCP/MOD1	bituminous based but differently activated compared to F300/F400	63.5	1.17	0.754	0.973	(Kommineni et al. 2003)
Centaur	catalytically enhanced coal	7.52	n.d.	0.336	n.d.	(Kommineni et al. 2003)
PCB	coconut-shell based coal	4.94	9.08	0.864	0.744	(Kommineni et al. 2003)
Alltech 5776	coconut-shell charcoal	4.26	1.31	0.894	1.11	(Kommineni et al. 2003)
Ambersorb 600	hydrophobic carbonaceous resin	10.2	4.09	0.669	0.866	(Kommineni et al. 2003)
Ambersorb 572	hydrophilic carbonaceous resin	4.92	2.17	0.93	1.07	(Kommineni et al. 2003)
Filtrasorb F400	contaminated and spiked groundwater	n.d.	0.96..1.27	n.d.	1.00..1.18	(Fleming et al. 1996)
CSC	coconut-shell carbon – contaminated and spiked groundwater	n.d.	0.31..2.57	n.d.	0.82..0.93	(Fleming et al. 1996)
Ambersorb 572	hydrophilic carbonaceous resin – contaminated and spiked groundwater	n.d.	9.65..28.37	n.d.	1.17..2.25	(Fleming et al. 1996)
Ambersorb 563	carbonaceous resin – contaminated and spiked groundwater	n.d.	2.09..7.35	n.d.	0.98..1.87	(Fleming et al. 1996)

GW: groundwater; n.d.: not determined

Alternative Treatment Technologies

Membrane Treatment

Reverse Osmosis. The NDMA molecule is uncharged. Thus, it is quite unlikely that it can be removed by reverse osmosis (Mitch et al. 2003a; Jobb et al. 1994). In experiments with an initial NDMA concentration of approximately 200 ng/L, Soroushian et al. (2001), however, found a removal of 12% for NDMA by reverse osmosis with cellulose acetate membranes. The removal was enhanced to 40% when thin film composite membranes were used.

Micro-, Ultra- and Nanofiltration. Membrane filtration for NDMA removal has so far not been investigated in more detail. However, experience shows that micro- or ultrafiltration is mainly dependent on the size of the substance to be removed. Due to its small size, NDMA provides no favorable properties concerning its removal from the water phase by membrane filtration. In nanofiltration units, however, removal is not only based on sieving effects but also on adsorptive characteristics and affinity of the membrane material towards NDMA. Thus, depending on the membrane material, nanofiltration may offer some potential for NDMA removal, even though economical considerations may prohibit the large scale implementation of this process in drinking water treatment.

Volatilization

Although NDMA exhibits a rather high vapor pressure, its concurrently high water solubility results in a low Henry coefficient (Table 6.6). Thus, NDMA removal by volatilization from aqueous solutions is rather improbable, and air stripping treatment during drinking water production will show almost no reduction in concentration (Mitch et al. 2003a). Fleming et al. (1996) confirmed these theoretical considerations by preliminary experiments and found no reduction in NDMA concentration due to volatilization.

Table 6.6
Physical-chemical properties of NDMA

k_{OH}	3.3×10^8 L mol^{-1} s^{-1}
M_{NDMA}	74.08 g/mol
boiling point 760 mm Hg	151-154°C
melting point (est.)	-50°C
vapor pressure (20°C)	1080 Pa 2.7 mm Hg
density (20°C)	1.0048 g/L
vapor density (25°C)	2.56 g/L
solubility	3.978 mg/L
Henry's law coefficient (est.)	3.34 Pa m^3 mol^{-1}
log K_{OW}	-0.57
log K_{OC}	1.41

Source: World Health Organization (2002)

Hydrolysis

Duré et al. (1978) investigated the behavior of different nitrosamines during hydrolysis. This reaction is initiated by cleavage of the covalent N-N bond by water.

$$(CH_3)_2N\text{-}NO + H_2O \rightarrow (CH_3)_2NH + HNO_2 \quad (\text{at pH} < 7)$$

The hydrolysis of nitrosamines follows first order kinetics. The substances investigated were NDMA, NDEA, NDPA, NPYR, and NPIP. Determined half-lives ranged from 45 to 700 weeks depending on the substance and pH value. The reaction rate constant for hydrolysis is dependent on the pH value. While the rate constant $k_{hydr.}$ is for all compounds almost constant at pH values above 4.5, a decisive increase of $k_{hydr.}$ is observed at pH values below 4.5. In acidic surroundings the nitrosamines are protonated resulting in an additional polarization of the molecule. This enables the subsequent attack of a water molecule by weakening of the N-N bond. As a result, $H_2^+NO_2$ (finally converted to nitrate) and the parent amine are formed. Under neutral to basic conditions the additional polarization does not occur and the reactivity of the nitrosamines towards water molecules is significantly reduced. At pH values occurring in the aquatic environment, half-lives for the nitrosamines under investigation range between 2.4 and 8.8 years depending on the compound. It can thus be concluded that hydrolysis has no major impact of the environmental fate of N-nitrosamines (Duré et al. 1978). In laboratory-scale experiments lasting 32 days, also Saunders and Mosier (1980) found no hydrolysis of NDPA at pH values of 3, 6, and 9.

Fleming et al. (1996) confirmed the acidic conversion of NDMA into dimethylamine and investigated if nitrosamine hydrolysis can be catalytically enhanced by Y-zeolite, silica, and Rohm-and-Haas Amberlyst® 15. Even though about 15 to 20% of the NDMA was removed in these experiments, the authors concluded that even catalytically enhanced hydrolysis is no option for eliminating NDMA effectively.

Removal with granular iron and nickel-enhanced iron

Gui et al. (2000) and Odziemkowski et al. (2000) investigated the use of zero-valent granular iron (Fe) and nickel-enhanced iron (Ni/Fe) as an electron source for reductively degrading NDMA in laboratory batch and column tests. NDMA solutions prepared in distilled water without deoxygenation were passed through columns packed with either granular iron or nickel-enhanced iron at concentrations ranging from 5 to 28 mg/L. At 5 mg/L NDMA, an $83.0 \pm 4.8\%$ removal of NDMA was found in the iron column with an average pseudo-first order rate constant of $(5.7 \pm 0.7) \times 10^{-2}\ h^{-1}$ and with an average half-life of 13 ± 2 hours. In the Ni/Fe column, the decomposition of NDMA was significantly enhanced (340 times). The reaction followed pseudo-first order for 273 pore volumes, and then changed to zero-order and mixed-order kinetics (Gui et al. 2000). In either case, DMA and ammonium were detected as NDMA transformation products. Neither UDMH nor N_2O were detected. Batch experiments showed that UDMH was not affected by Fe or Fe/Ni, ruling out that substance as an intermediate (Gui et al. 2000). The much faster reduction rate of NDMA observed with Ni/Fe than with Fe can be explained by a catalytic hydrogenation reaction mechanism since Ni is a stronger hydrogenation catalyst than Fe (Odziemkowski et al. 2000). During catalytic dehydrogenation the charge transfer takes place between Ni/Fe and water molecules forming adsorbed atomic hydrogen (FeH_{ads} or NiH_{ads}). The latter is then responsible for the hydrogenation of NDMA. This is supported by mechanistic considerations which show that the redox potential of Fe (-0.102 to -0.52 V) is not negative enough for the electro-reduction of NDMA (-1,31 to -1,38 V) but for hydrogen evolution forming adsorbed atomic hydrogen as intermediate (FeH_{ads}) (Odziemkowski et al. 2000). During the experiments, a loss in reactivity in the Fe/Ni column was observed due to the adsorption of reaction by-products (DMA, ammonium), surface coverage by corrosion products, or a loss of galvanic coupling. The reactivity could, however, be partially restored by flushing the column with acidic solution with only a small amount of Fe and Ni washed out (Gui et al. 2000).

Phytoremediation

Although NDMA is not present in plants under natural conditions, it can be taken up from the growth medium. Lettuce and spinach plants absorb NDMA from sand, soil, and water. After exposure for 2 days to concentrations ranging from 10 to 100 mg NDMA/kg wet weight, 3.25 and 0.8 % NDMA were taken up from the growth medium by lettuce and spinach plants, respectively (Dean-Raymond and Alexander 1976). The absorbed NDMA disappeared with time, in agreement with results on nitrosamine uptake by Lepidium sativum (Dean-Raymond and Alexander 1976). Stiborová et al. (2000) reported that the microsomal fraction from tulip bulbs (Tulipa fosteriana, L.) contains cytochrome P450 and peroxidase enzymes catalyzing the NADPH-dependent oxidation of NDMA. It has also been reported that cereals can absorb nitrosamines (Dean-Raymond and Alexander 1976). Thus, bioremediation might have some potential for NDMA removal that has so far not been evaluated in more detail.

LABORATORY-SCALE EXPERIMENTS

Biodegradation of Nitrosamines

The removal of nitrosamines by riverbank filtration and artificial groundwater recharge has so far not been investigated in more detail. In general, the fate of organic micropollutants during infiltration and underground passage is mainly determined by adsorption mechanisms and microbial degradation. Literature results clearly demonstrate that NDMA can be degraded by microorganisms under certain conditions, while sorption processes to sediments or soil turned out to be rather insignificant. Most investigations so far were conducted via incubation experiments with soils using relatively high NDMA concentrations.

Therefore, the objective of this part was to investigate the biodegradation of NDMA, NEMA, NDEA, NDPA, NDBA, NPIP, NPYR, NMOR, and NDcHxA using an established test filter device to simulate the aerobic microbial degradation processes taking place during riverbank filtration, artificial groundwater recharge or in biofiltration units. The method allowed studying the aerobic biodegradability of nitrosamines at concentration levels relevant for drinking water (100 ng/L), while sorption effects to sediments or soils were consciously avoided. The surface water used as medium and for establishment of the microorganisms on the test filter carrier material was characterized as follows: medium wastewater impact, no detectable nitrosamines (< 0.3 ng/L), 2.0 mg/L TOC, pH 8.1, 45 mS/m conductivity, 6.0 mg/L nitrate, 0.02 mg/L ammonia, 62 mg/L chloride, and 28 mg/L sulfate.

The microbial degradation of each nitrosamine was assessed from the time-dependent course of the concentration levels (c/c_0). The degradation behavior of the nitrosamines in the test filter is shown in Figures 6.9 and 6.10.

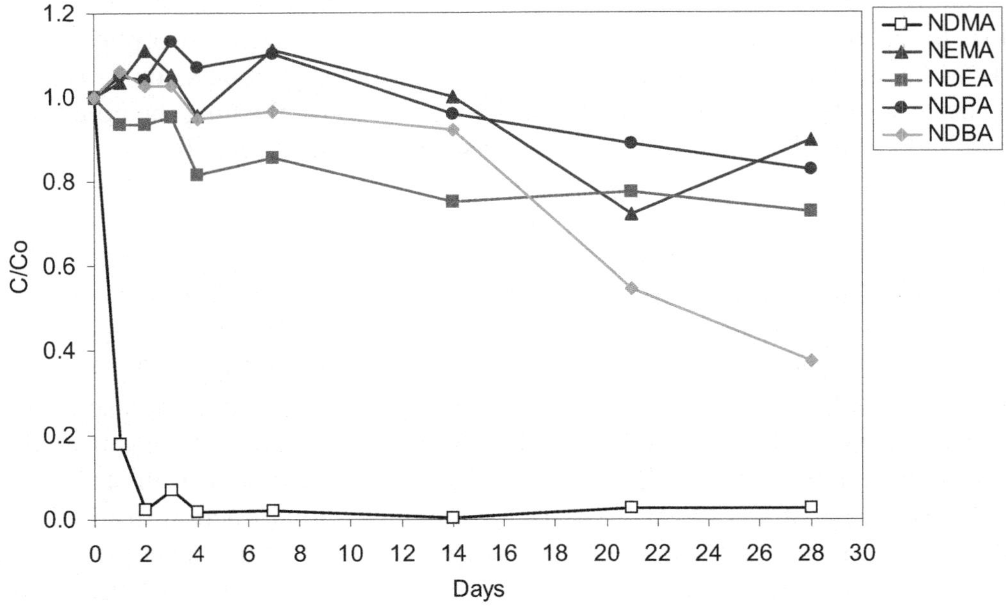

Conditions: c_0 = 100 ng/L

Figure 6.9 Microbial degradation of aliphatic nitrosamines in a test filter

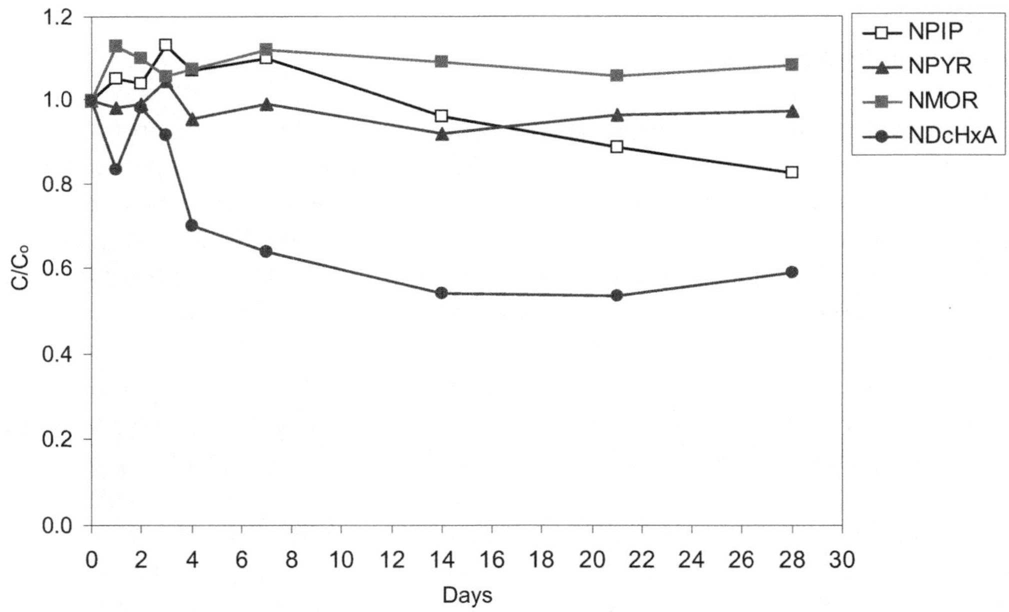

Conditions: c_0 = 100 ng/L

Figure 6.10 Microbial degradation of alicyclic nitrosamines in a test filter

As can be seen from Figure 6.9, the concentrations of NDMA decreased quite rapidly, indicating a fast microbial degradation. Compared to NDMA, the concentrations of the other nitrosamines decreased much more slow, suggesting that their degradation proceeds retarded or is limited. According to the results of these experiments, aerobic biodegradability of the nitrosamines decreases in the order NDMA >> NDBA > NDcHxA > NDEA > NPIP, NDPA, NEMA > NPYR > NMOR. Since the surface water used was not contaminated with NDMA prior to the experiments, the fast degradation of NDMA implies that microbial adaptation processes seem to be of minor relevance for NDMA removal by biofiltration.

The good elimination of NDMA during biofiltration is surprising at first sight, since published studies reported on rather long half-lives. Purification processes taking place in biofiltration units are similar to the self-cleaning properties found in surface waters or contaminated soils, but proceed in the biofilm established by continuous filtration much more intensively. Due to the continuously provided supply of substrates, the microbial density in biofiltration units is typically higher than in soils. Consequently, NDMA biodegradation should also be enhanced when compared to incubation experiments with soil. The microbial degradation of organic substrates is closely linked to redox reactions catalyzed by the micro-organisms. The energy gain is highest for aerobic respiration, so that during microbial degradation of organic substances in water dissolved oxygen is consumed first. Varying micro-organisms, that are adapted to the particular milieu conditions, can cause, by successive depletion of alternative, less efficient electron acceptors (NO_3^-, Mn(IV), Fe(III), SO_4^{2-}), continued degradation of organic compounds in a sequence of reduction steps, with redox conditions changing to anaerobic conditions. Elimination rates and metabolic pathways may differ substantially in different redox milieus, so that the results presented here for the aerobic case are not necessarily transferable to an anoxic or anaerobic biofiltration process. Recently, Hoppe et al. (2006) reported on the facility with which microorganisms removed NDMA, NDEA, NDPA, NDBA, NPIP and NPYR in saturated anoxic soil columns continuously fed with a spiked groundwater. Comparing these results with the results from the testfilter device, NDEA, NDPA, NDBA and NPYR seem to be better degradable under anoxic than under aerobic conditions. Neither the testfilter experiments presented here nor the investigation by Hoppe et al., however, examined the metabolites formed during the biodegradation process. Possibly, the reduction under anoxic conditions is only related to a simple reduction of the nitroso-group into the corresponding hydrazine, as was reported in the literature for mammalian cells, while the oxidative route may cause a complete mineralization.

Oxidation

The objective of this part was to evaluate various oxidants with regard to their ability to destruct NDMA. Furthermore, experiments were performed to compare NDMA removal using conventional ozonation or oxidation by the AOP O_3/H_2O_2.

Rate Constants of Reactions of NDMA with Various Oxidants

NDMA is strongly resistant to oxidation because of the electron withdrawing nitroso (-NO) group. Table 6.7 summarizes the determined second-order rate constants for reactions of NDMA with several oxidants. The rate constants are very low except for the reaction with $^{\bullet}OH$, which is known as a highly reactive and non-selective oxidant.

Table 6.7
Rate constants for reactions of NDMA with oxidants

Oxidant	Rate constant ($M^{-1}s^{-1}$)
ClO_2	$< 10^{-3}$
Fe(VI)	1.1 (at pH 7)
O_3	5.2×10^{-2}
$^{\bullet}OH$	4.5×10^{8}

Oxidation of NDMA in Natural Waters by Ozonation

It has been demonstrated that ozonation is an effective technology for the oxidation of micropollutants in water. Ozone can either react directly with organic compounds, or decompose to OH radicals ($^{\bullet}OH$), which also react with the target compounds (von Gunten 2003). To enhance oxidation of ozone-resistant compounds, hydrogen peroxide can be added which enhances $^{\bullet}OH$ production by accelerating ozone decomposition (Acero and von Gunten 2001).

To simulate the NDMA oxidation by ozonation under real treatment conditions, experiments were performed in natural lake waters (lake waters A and B). Figure 6.11 shows the NDMA oxidation by conventional ozonation and the AOP O_3/H_2O_2 with various doses of oxidants in lake water A. About 80% degradation of NDMA by the AOP O_3/H_2O_2 required 320 μM of ozone. The conventional ozonation with 160 μM led to only 25% NDMA oxidation, whereas the AOP O_3/H_2O_2 with the same ozone dose resulted in 55% NDMA oxidation.

To improve the efficiency of NDMA oxidation by the AOP O_3/H_2O_2 in lake water A, 160 μM ozone was split into doses of 2 × 80 μM, 4 × 40 μM, and 8 × 20 μM, respectively, and injected into the solution at regular intervals during the reaction time. The ratio $[O_3]_0/[H_2O_2]_0$ of 2 was maintained constant for all experiments, and H_2O_2 was added immediately before ozone injection. This split injection of oxidants can minimize $^{\bullet}OH$ scavenging by ozone by maintaining low concentration of ozone during the whole reaction. Figure 6.12 shows the influence of split injection of ozone on NDMA oxidation by the AOP O_3/H_2O_2 in lake water. The degree of NDMA oxidation increased with increasing the number of ozone injections. As shown in Figure 6.12, a single injection of 160 μM ozone yielded 55% of NDMA oxidation, whereas for 8 injections of 20 μM ozone, 75% was achieved. This is similar to the degree of oxidation by a single initial injection of 320 μM ozone (78%).

Similar experiments were performed using another lake water (lake water B), which contains a much higher concentration of organic and inorganic matrix components than the one used for the previous experiments. Figure 6.13 shows the effect of oxidant dose on NDMA oxidation in lake water B using conventional ozonation and the AOP O_3/H_2O_2. Comparing the results of Figures 6.11 and 6.13 for the same ozone dose, NDMA removal efficiencies in lake water B were obviously lower than in lake water A due to the higher $^{\bullet}OH$ scavenging rate of lake water B. Conventional ozonation with up to 160 μM ozone led to less than 10% NDMA oxidation, and about 320 μM ozone ($[O_3]_0/[H_2O_2]_0 = 2{:}1$) was required for 50% NDMA oxidation. A split injection of oxidants did not significantly improve the NDMA oxidation efficiency in lake water B (Figure 6.14).

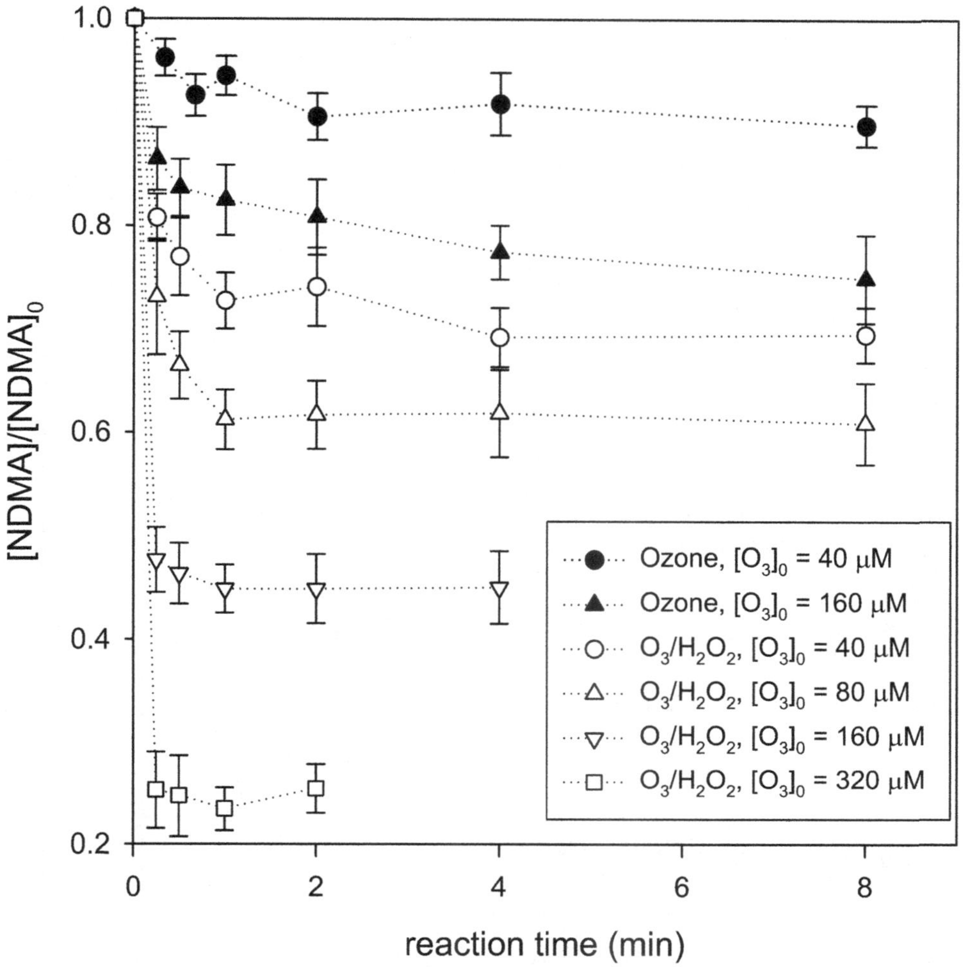

Conditions: $[NDMA]_0 = 1$ μM, ratio of $[O_3]_0/[H_2O_2]_0 = 2$, pH = 7.9

Figure 6.11 Influence of ozone concentration on NDMA oxidation by conventional ozonation and the AOP O_3/H_2O_2 in lake water A

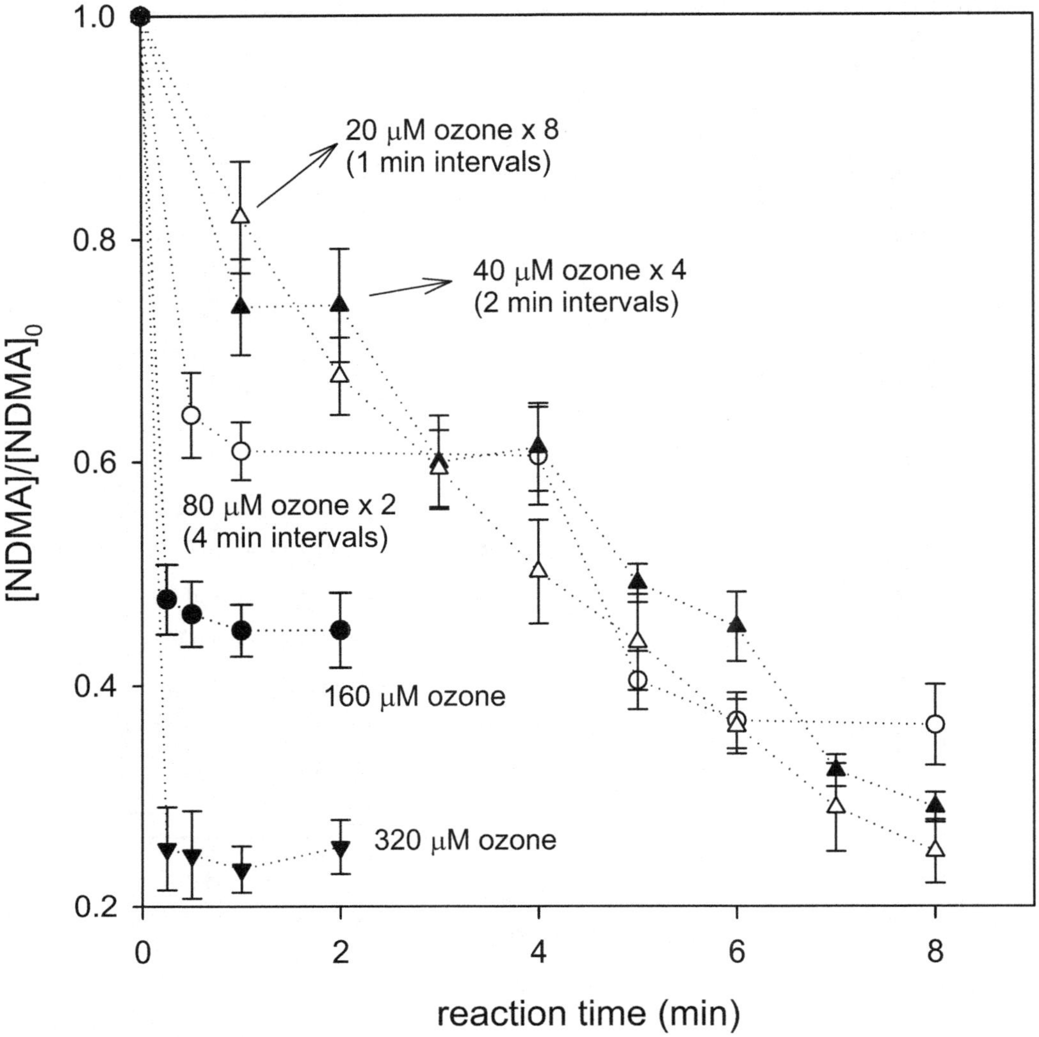

Conditions: $[NDMA]_0 = 1\ \mu M$, ratio of $[O_3]_0/[H_2O_2]_0 = 2$, pH = 7.9

Figure 6.12 Influence of split injection of oxidants on NDMA oxidation by the AOP O_3/H_2O_2 in lake water A

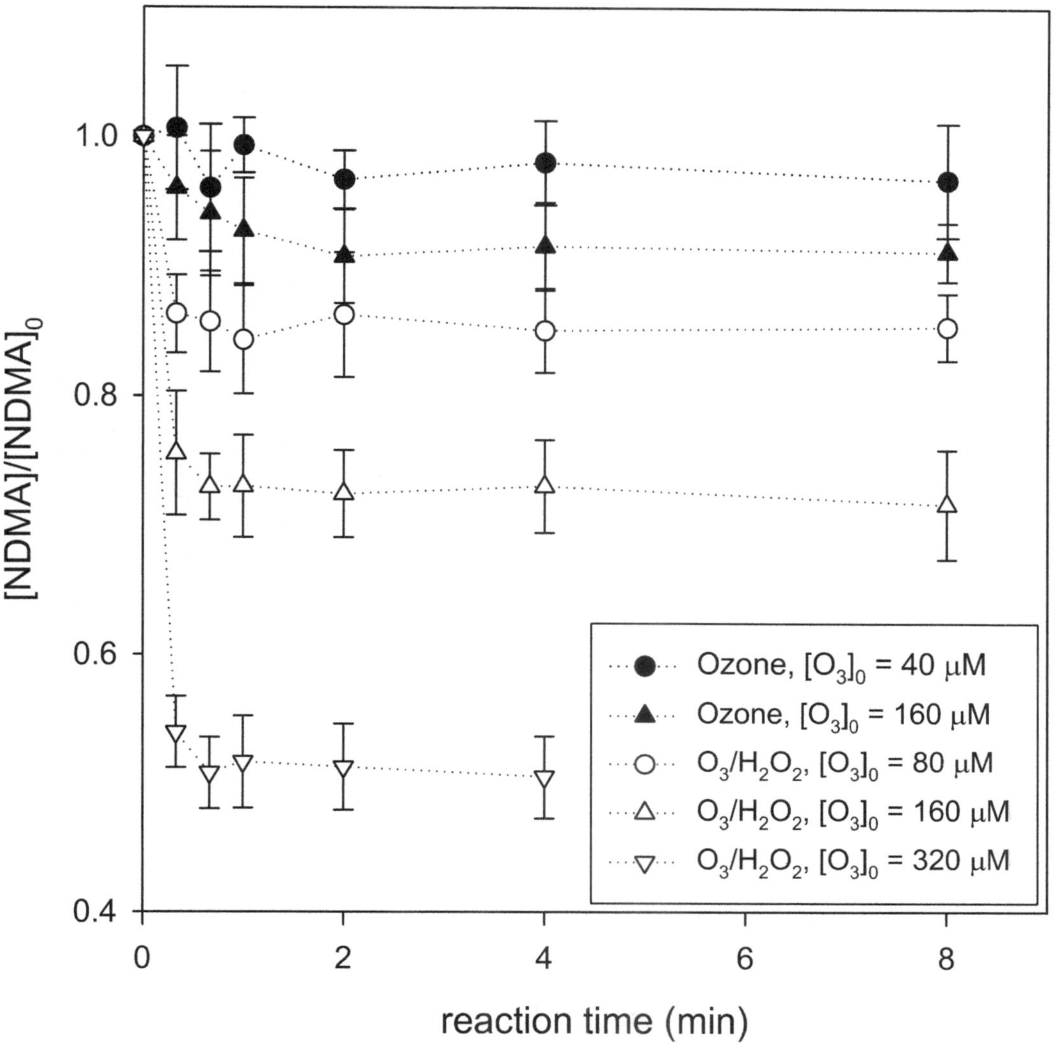

Conditions: $[NDMA]_0 = 1$ μM, ratio of $[O_3]_0/[H_2O_2]_0 = 2$, pH = 8.1

Figure 6.13 Influence of ozone concentration on NDMA oxidation by conventional ozonation and the AOP O_3/H_2O_2 in lake water B

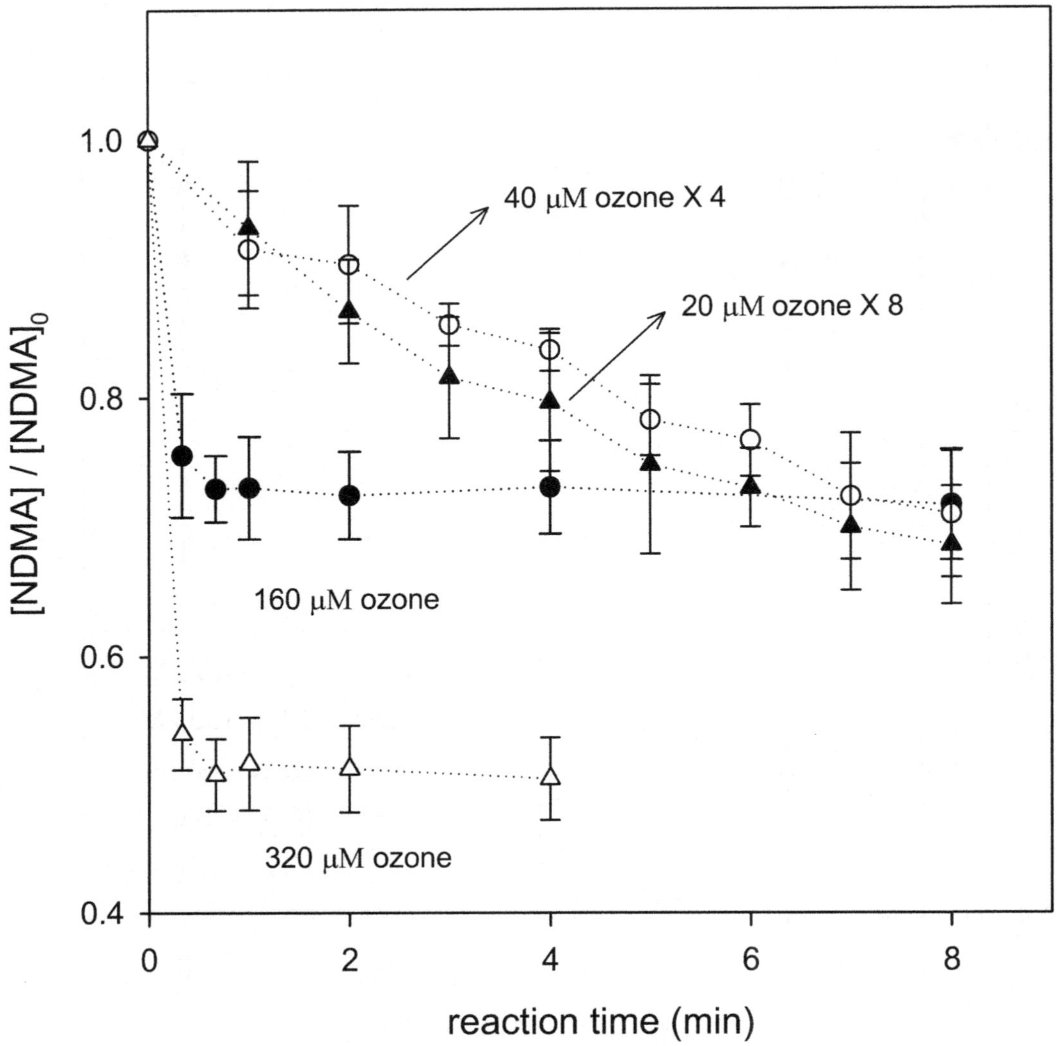

Conditions: $[NDMA]_0$ = 1 μM, ratio of $[O_3]_0/[H_2O_2]_0 = 2$, pH = 8.1

Figure 6.14 Influence of split injection of oxidants on NDMA oxidation by the AOP O_3/H_2O_2 in lake water B

CHAPTER 7
FULL-SCALE TESTING IN WATER UTILITIES

INTRODUCTION

In order to verify the results from laboratory-scale experiments under real-world conditions, full-scale tests were performed. Samples from five U.S. utilities, one Australian utility, and three German utilities were obtained and analyzed for nitrosamines. Additionally the nitrosamine formation potential was determined for each sample using the NFP test described in Chapter 2 (pH = 7.0, 0.4 mM chloramine, 168 h, 20 °C). In the following sections the results of these tests will be presented and discussed for each utility. Special focus will be put on a comparison of the results of the full-scale tests to the outcomes of the laboratory-based experiments.

UTILITY A

Investigations at utility A were focused on the applied riverbank filtration process only. The corresponding riverbank filtration site is shown in Figure 7.1 in the cross section profile. Extraction wells were constructed at a distance of about 170 m from the bank in parallel to the river that is impacted by industrial and municipal wastewater. The wells extract a mixture of bank-filtered water and native land-sided groundwater from the sandy-gravely, about 15 m thick terrace deposits of the river. The aquifer is characterized by a pronounced anisotropy of hydraulic conductivities. Thereby, the lower layer shows an at least six-fold higher permeability than the upper layer. Due to an almost impermeable layer at the river bank, the water infiltrates in particular at the river bottom in the middle of the river. The majority of the water collected from the production wells originates from the lower aquifer layer and the produced raw water contains about 60% riverbank filtrate.

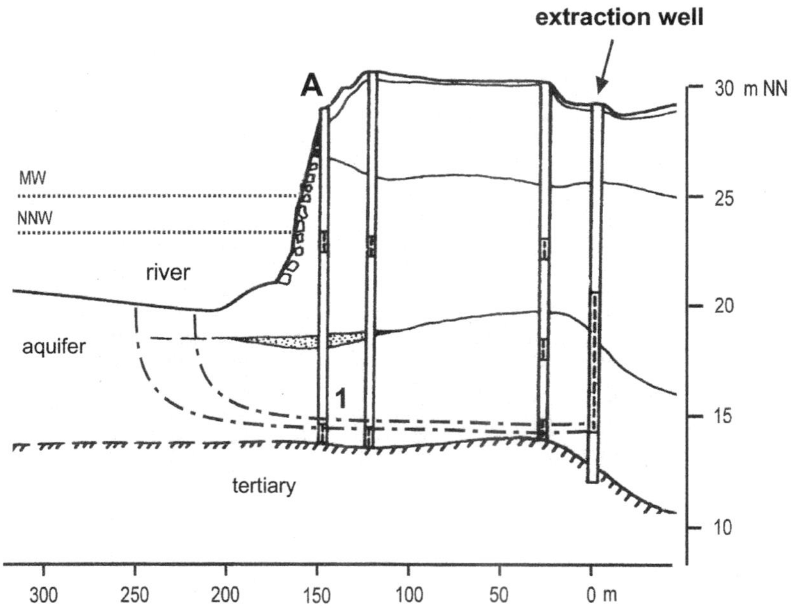

Figure 7.1 Cross section profile of the investigated riverbank filtration site of utility A

Between well gallery and bank line a number of observation wells are established. Investigations presented here were performed at the singly filtered level A1 with deep filter position that provides pure riverbank filtrate without any land-sided groundwater impact. The residence time of the infiltrated water in the underground until this sampling point is reached is between 7 and 20 days. According to the year-long experience and observations of the operator, redox conditions along the aquifer are aerobic. Although the oxygen dissolved in the surface water (median: 9.2 mg/L) is significantly reduced during underground passage to sampling point A1 (median: 1.1 mg/L), the redox milieu within the subsurface is characterized throughout the year by aerobic conditions, with no nitrate reduction taking place. Parameters like conductivity, chloride, and sulfate are comparable in the river and the bank filtrate and prove that sampling site A1 is not affected by land-sided groundwater.

Measurements of aliphatic amines and NDMA were performed several times for the river surface water and at the same time for the respective bank filtered water in sampling well A1 over a time period of several months in order to get reliable data on the removal of these compounds during riverbank filtration. The data sets obtained for aliphatic amines are summarized in Figures 7.2 and 7.3 in the box plot presentation. As can be seen from the surface water data in Figure 7.2, amine concentrations in this river typically amount to 30-400 ng/L per single substance. On an average, highest concentrations were found for DMA, DEA, and morpholine. Pyrrolidine concentrations were comparably low, with levels rarely exceeding 50 ng/L, and piperidine was never detected above its detection limit of 15 ng/L. Comparing amine levels in the surface water with those in the riverbank filtrate revealed that concentrations of all amines can be lowered by the riverbank filtration process, even though a complete removal could not be observed for any compound. However, the average elimination efficiency was 76% for DMA, 66% for DEA, 52% for pyrrolidine and 80% for morpholine, indicating that relevant secondary nitrosamine precursor amines are significantly reduced by the riverbank filtration process.

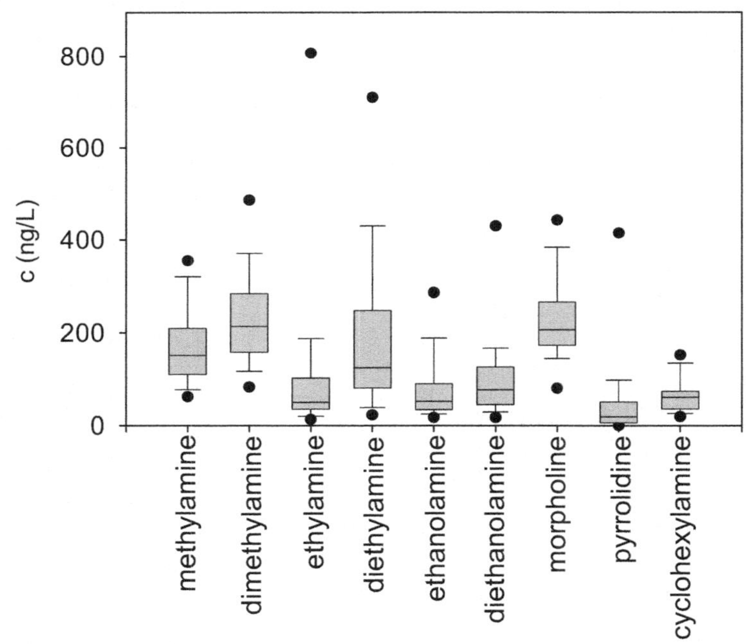

Data from 09/2005-05/2006, n = 35

Figure 7.2 Concentrations of aliphatic amines in the river source water used by utility A for riverbank filtration

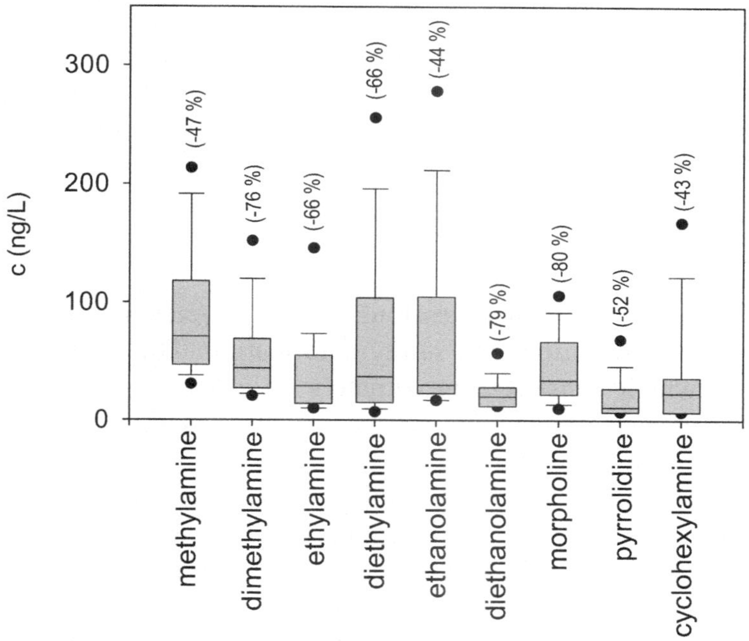

Sampling point A1 with no dilution by natural land-sided groundwater, in brackets average removal efficiency; data from 09/2005-05/2006, n = 35

Figure 7.3 Concentrations of aliphatic amines in the riverbank filtrate of utility A

Since the river at this utility was also contaminated with NDMA at concentrations between the detection limit of 0.3 ng/L and 10 ng/L, the removal efficiency of NDMA could also be investigated at this site. Levels of NDMA in the surface water and the corresponding bank filtrate are shown in Figure 7.4.

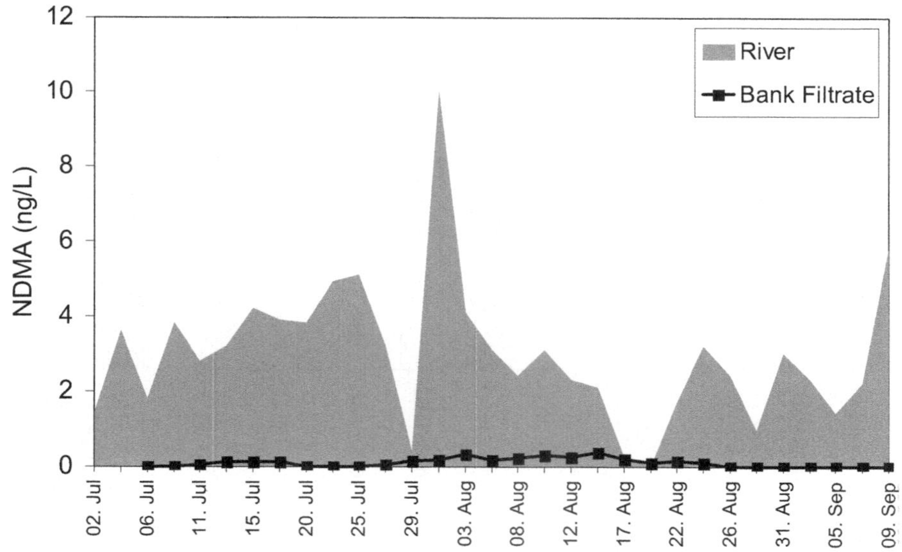

Sampling point A1; data from 2005, n = 31

Figure 7.4 NDMA concentrations in a river and the corresponding bank filtrate

The data clearly confirm the findings from the biologically active test filter device run at aerobic redox conditions. NDMA is well eliminated (> 90%) during the aerobic riverbank filtration process. Based on the results of the testfilter device and due to the fact that sorption of NDMA to sediments and soil was shown to be insignificant in earlier studies, the fate of NDMA during riverbank filtration should be mainly determined by biological transformations.

UTILITY B

Water treatment at utility B involves four distinct processing steps: riverbank filtration, ozonation, activated carbon filtration, and safety chlorination. The wells of the waterworks extract riverbank filtrate which consists approximately of a quarter of natural groundwater and three quarters of river water seepage. The passage of the surface water through the bank and layers of sand and gravel, up to 30 meters thick, takes about 30-60 days. During further processing, this raw water flows through a contact tank and ozone is added at a concentration of 0.5 mg/L. The ozone treated water is retained in an intermediate tank for some 30 minutes reaction time. The third step is filtration with activated carbon. The water is pumped into a steel tank filled with two layers of carbon. The upper layer is an old, 1.5 m deep reactivated carbon type; the second layer, 2.5-m deep, consists of fresh activated carbon. Finally, safety chlorination is performed with 0.06 mg/L chlorine dioxide.

The water from utility B was sampled along the treatment process at three locations: surface water, raw riverbank filtrate, and activated carbon effluent. Altogether, eight sets of single grab samples were analyzed for nitrosamine precursor levels via the NFP test. The results are summarized in Figure 7.5. More general water quality data in the sampled waters of this utility are given in Table 7.1.

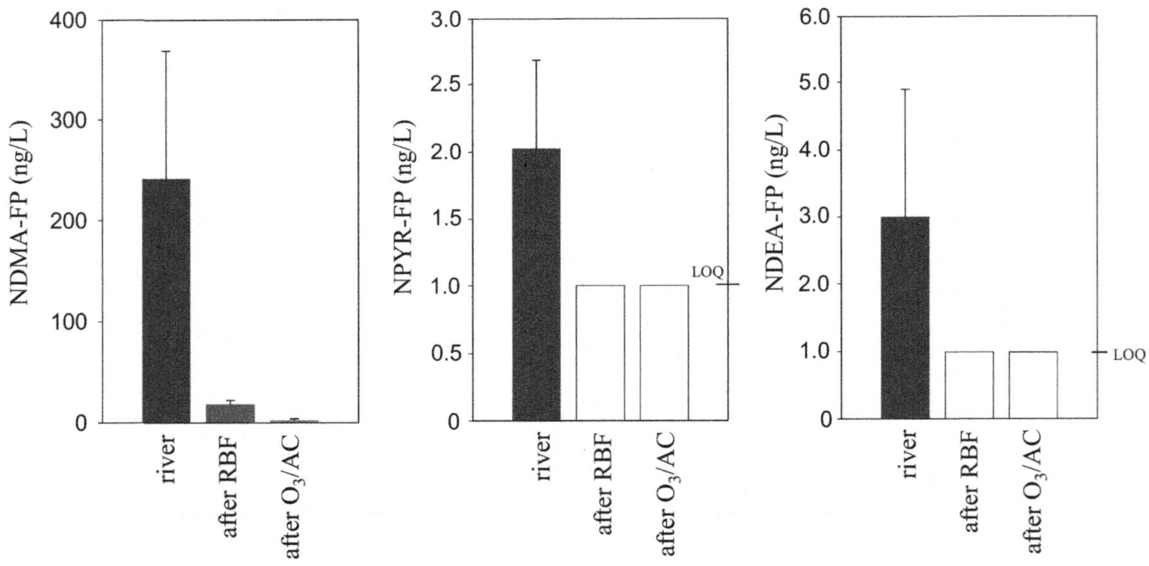

Conditions: NFP test (0.4 mM chloramine, pH 7, 20 °C, 168 h), RBF – riverbank filtration, AC – activated carbon; data from 2004-2006, n = 8

Figure 7.5 Nitrosamine precursor levels determined by the NFP along the water treatment process at utility B

Table 7.1
Physical-chemical characteristics of water samples from utility B

Water sample (sampling date)	pH	Alkalinity (mg/L)	Ammonia (mg/L N)	Nitrate (mg/L N)	Nitrite (mg/L N)	TOC (mg/L)	UV_{254} (m^{-1})
River (04.07.2006)	8.06	128	0.04	9.5	0.02	2.5	5.7
Riverbank filtrate (04.07.2006)	7.85	163	<0.01	9.9	<0.01	1.0	1.5
Activated carbon effluent (04.07.2006)	7.78	162	<0.01	9.8	<0.01	0.41	0.3

Based on the results presented in Figure 7.5, it becomes obvious that the NDMA formation potential exceeded that of NPYR and NDEA approximately by two orders of magnitude. Although the sample sets under investigation were not timed to account for the lag during bank transport, the distinct reduction in the NFP clearly indicates that riverbank filtration is capable of removing precursors of NDMA, NPYR, and NDEA to a large extent, with removal efficiencies for the NDMA FP of 93%. NDMA precursor levels in the raw riverbank filtrate (mean: 18 ng/L) could be further reduced by about 90% via the combination of activated carbon treatment and ozonation. The average NDMA formation potential in the activated carbon effluent was 2.0 ng/L, compared to 230 ng/L in the river.

The results of these full-scale experiments confirm in part the results from the lab-scale studies where treatment steps based on activated carbon and ozonation were demonstrated to be effective in NDMA precursor removal. However, laboratory-based experiments with the biologically activated test filter device demonstrated that NDMA precursor compounds are rather persistent to biotransformation making their removal during riverbank filtration based on this process rather improbable. However, riverbank filtration also offers the possibility of compound removal by adsorption processes. Thus, the results from the laboratory-scale experiments and this full-scale study might indicate that NDMA precursors can be adsorbed by sediment or soil or might be associated with particles being filtered. In studies by Gerecke and Sedlak (2003), however, almost no NDMA precursors were associated with particles greater than 0.7 μm in surface waters, pointing at adsorption processes as being of central importance. The possible relevance of adsorption processes is further underlined by the fact that NDMA precursors turned out to be well removable by activated carbon treatment (at least after ozonation).

UTILITY C

Utility C processes its drinking water via artificial groundwater recharge of surface water followed by deacidification with sodium hydroxide and safety chlorination with chlorine dioxide. After pre-filtration through fine gravel filters the river water is recharged by slow sand filtration. The oxic infiltrated water is captured by a drain tile after an average residence time of three days in the aquifer. The investigations performed at this site aimed at studying the influence of artificial groundwater recharge. Therefore, four sets of samples were collected at two locations (river water and recharged groundwater) and were analyzed for nitrosamine precursor levels via the NFP test. Figure 7.6 shows the results of this investigation; Table 7.2 gives information on the more general water quality parameters at the sampling points of this utility.

Table 7.2
Physical-chemical characteristics of water samples from utility C

Water sample (sampling date)	pH	Alkalinity (mg/L)	Ammonia (mg/L N)	Nitrate (mg/L N)	Nitrite (mg/L N)	TOC (mg/L)	UV_{254} (m^{-1})
River (04.11.2004)	8.09	108	0.08	12.9	0.08	2.6	2.2
Recharged groundwater (04.11.2004)	7.71	111	<0.01	12.0	<0.01	0.93	1.3

The results confirm in principle the results of the investigations at utility B. Again, the NDMA-FP exceeds that of NPYR significantly (ca. 30fold). Levels of NDEA formed during the NFP test were always below the analytical LOQ of 1 ng/L. Similar to the observations made for the riverbank filtration process at utility B, the artificial groundwater recharge process at utility C substantially removed precursors of NDMA and NPYR. Average removal efficiencies by artificial groundwater recharge were about 88% for the NDMA-FP and > 63% for the NPYR-FP and were thus comparable to those observed for the riverbank filtration process at utility C, implying that residence times in the aquifer are of minor relevance. Since the surface water utilized by utility B is different from the one processed by utility C, the results indicate the general possibility of NDMA and NPYR precursor removal by treatment processes based on an underground passage.

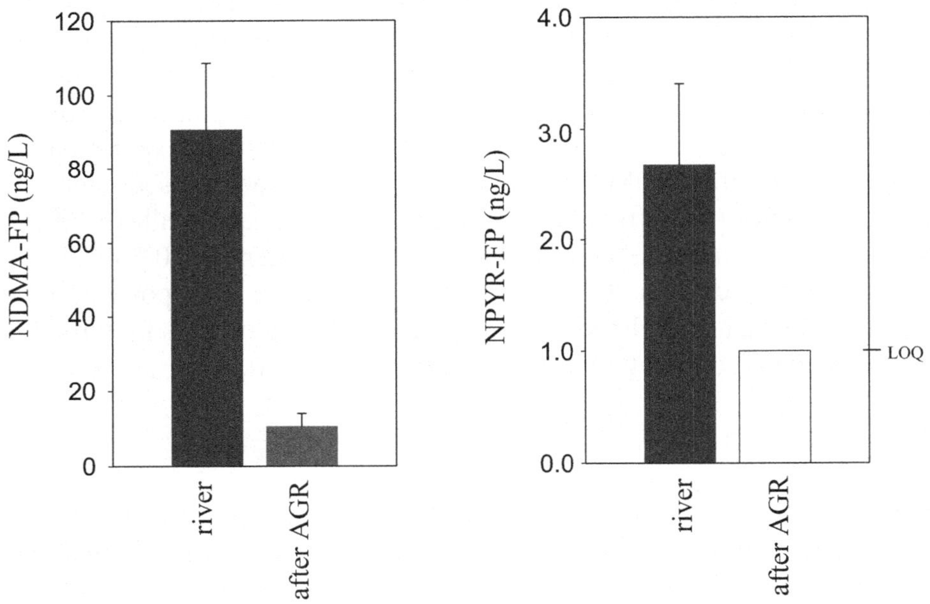

Conditions: NFP test (0.4 mM chloramine, pH 7, 20 °C, 168 h), AGR – artificial groundwater recharge; data from 2004-2006, n = 4

Figure 7.6 Nitrosamine precursor levels determined by the NFP along the water treatment process at utility C

UTILITY D

For utility D, water samples from two different water treatment plants (WTPs), both treating river water, were investigated. Both river sources are moderately impacted by wastewater discharges upstream of plant intakes. The raw water of WTP B is also influenced by a heavily wastewater impacted tributary which feeds into the river just upstream of the plant intake. The two plants are operated with conventional treatment and encompass the following sequence of processes: raw water basin, rapid mix (ferric chloride, lime (if necessary), and small chlorine dose), flocculation (with chlorine addition at mid-floc), sedimentation, filtration (chlorine added as water enters filters), ammonia, fluoride, and zinc orthophosphate addition at the filter effluent (distributed water has chloramine residual), clearwell, and distribution.

The main difference between WTP A and WTP B can be seen in the use of the backwash water of the filtration unit: plant A returns its backwash water to the head of the plant (i.e. into the raw water basin) whereas plant B sends it to waste. A treatment scheme of the two plants of utility D illustrating the addition of chemicals as well as the streams of water and the locations where the samples have been taken is given in Figure 7.7. Details on the operating conditions of WTP A and WTP B at the sampling date (01.03.2006) are summarized in Table 7.3. Tables 7.4, 7.5, and 7.6 summarize some water quality data for the samples under investigation as well as the results of nitrosamine analysis and NFP testing.

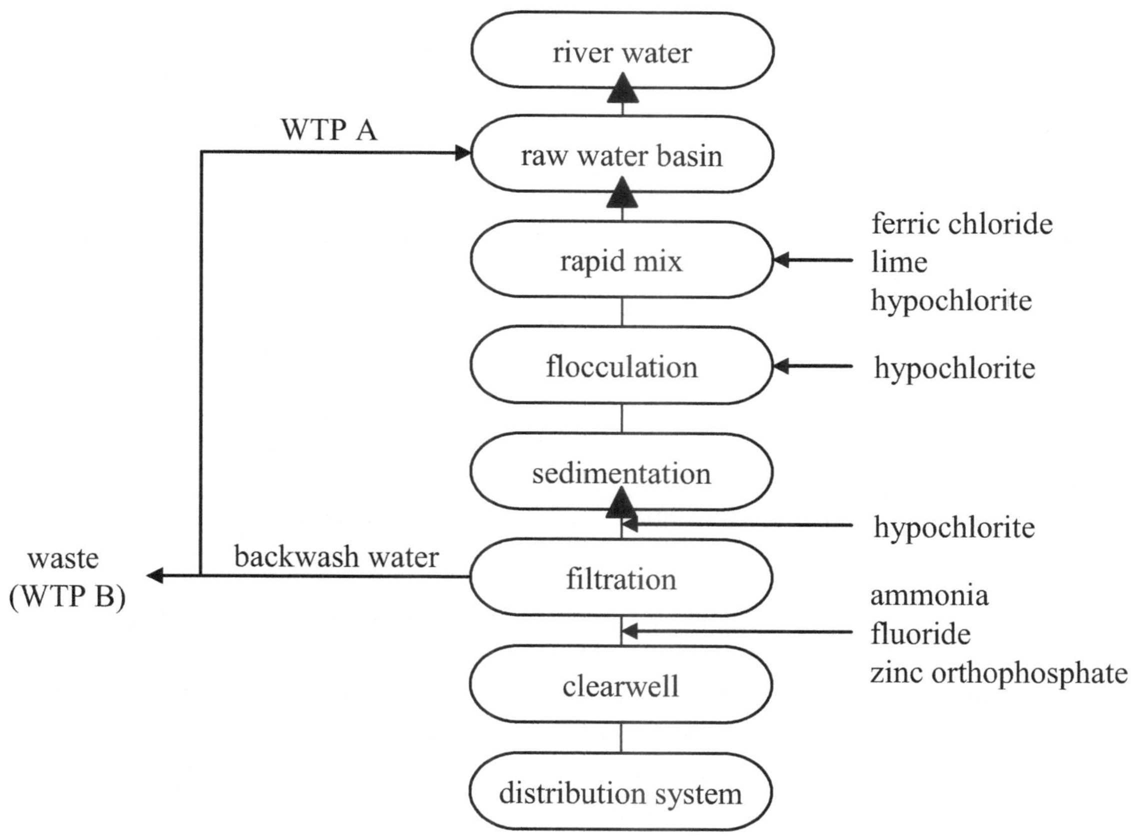

▲: sampling locations

Figure 7.7 Treatment scheme of WTPs A and B of utility D

Table 7.3
Operating conditions of WTP A and WTP B from utility D at 01.03.2006

Parameter	WTP A	WTP B
Intake chemical dosing	none	none
Pretreatment chemical dosing	16.8 mg/L ferric chloride 2.4 mg/L hypochlorite no lime	41.9 mg/L ferric chloride 1.9 mg/L hypochlorite no lime
Detention time between pretreatment and flocculation	2.5 min	3.5 min
Detention time for flocculation	78 min	58.7 min
Detention time for sedimentation	285.6 min	267.7 min

Source: Data provided by utility D

Table 7.4
Physical-chemical characteristics of water samples from utility D

Water sample (sampling date)	pH	Alkalinity (mg/L)	Ammonia (mg/L)	Nitrate (mg/L)	Nitrite (mg/L)	DOC (mg/L)	TOC (mg/L)	UV_{254} (m^{-1})
WTP A raw water (01.03.2006)	8.4	96	0.068	3.91	<0.05	1.52	1.82	6.4
WTP A raw water basin effluent (01.03.2006)	8.4	89	0.044	3.69	<0.05	1.47	1.78	5.1
WTP A clarifier effluent (01.03.2006)	6.5	43	0.049	3.68	<0.05	1.07	1.19	2.6
WTP B raw water (01.03.2006)	7.7	46	0.829	1.14	<0.05	1.36	1.59	6.6
WTP B raw water basin effluent (01.03.2006)	7.7	48	0.101	1.26	<0.05	1.40	1.62	6.6
WTP B clarifier effluent (01.03.2006)	6.8	33	0.051	1.30	<0.05	0.97	1.05	2.9

Source: Data provided by utility D; Alkalinity in mg/L $CaCO_3$, Concentrations of ammonia, nitrite and nitrate given as mg/L NH_4^+, mg/L NO_2^-, and mg/L NO_3^-

As it can be seen from the data in Table 7.5, no nitrosamines could be detected in the original water samples. After chloramination under the conditions of the NFP test, NDMA as well as NPYR (N-nitrosopyrrolidine) could be detected in the raw waters of both treatment plants. For WTP B, the NFP remains unchanged in the raw water basin, whereas in WTP A there is an increase which is probably due to the impact of recycled backwash water. Only in plant A the backwash water is recycled whereas plant B sends it to waste (see also Figure 7.7). However, as the raw water basin detention time for plant A is 12 to 18 hours, the differences for a pair of influent and effluent samples could also stem in part from changes in river water quality over that time span (the samples were not staggered to account for the detention lag).

Table 7.5
Nitrosamine concentrations in original water samples from utility D

Water sample	NDMA (ng/L)	NEMA (ng/L)	NDEA (ng/L)	NDPA (ng/L)	NDBA (ng/L)	NPIP (ng/L)	NPYR (ng/L)	NMOR (ng/L)
WTP A raw water	<1	<2	<2	<1	<1	<1	<1	<1
WTP A raw water basin effluent	<1	<2	<2	<1	<1	<1	<1	<1
WTP A clarifier effluent	<1	<2	<2	<1	<1	<1	<1	<1
WTP B raw water	<1	<2	<2	<1	<1	<1	<1	<1
WTP B raw water basin effluent	<1	<2	<2	<1	<1	<1	<1	<1
WTP B clarifier effluent	<1	<2	<2	<1	<1	<1	<1	<1

Table 7.6
NFP of water samples from utility D

Water sample	NDMA (ng/L)	NEMA (ng/L)	NDEA (ng/L)	NDPA (ng/L)	NDBA (ng/L)	NPIP (ng/L)	NPYR (ng/L)	NMOR (ng/L)
WTP A raw water	39	<2	<2	<1	<1	<1	1.0	<1
WTP A raw water basin effluent	71	<2	<2	<1	<1	<1	1.6	<1
WTP A clarifier effluent	9.1	<2	<2	<1	<1	<1	<1	<1
WTP B raw water	85	<2	<2	<1	<1	<1	13	<1
WTP B raw water basin effluent	83	<2	<2	<1	<1	<1	7.6	<1
WTP B clarifier effluent	11	<2	<2	<1	<1	<1	<1	<1

In both treatment plants of utility D, the applied water treatment processes result in a significant elimination of precursors and the NFP values in the clarifier effluent samples are an order of magnitude lower than in raw water basin effluents. The reduction in NFP in the course of the treatment process might be the result of either the coagulation/flocculation step or the addition of chlorine. Previous laboratory-scale experiments, however, demonstrated no removal of NDMA precursors by flocculation with polyaluminum chloride (PAC) and iron chloride, so that the present findings are more likely attributable to the chlorination state of organic nitrogen precursors. This result is an excellent confirmation of the results from the laboratory-scale experiments on oxidation of possible nitrosamine precursors, where it could be proven that ozonation or chlorination leads to a significant reduction in NFP.

Table 7.7
Removal efficiencies for different parameters in WTPs A and B from utility D

	Removal efficiency (%)			
	DOC	UV_{254}	NFP(NDMA)	NFP(NPYR)
WTP A	27	49	87	>38
WTP B	31	56	87	>87

In order to get further insight in the nature of the nitrosamine precursor, Table 7.7 gives an overview on the removal efficiencies of the two treatment plants of utility D with respect to the parameters DOC, UV_{254}, and formation potential for NDMA and NPYR. The data were calculated from the respective values in the raw water basin effluent and in the clarifier effluent. For calculation of the NPYR FP removal efficiencies, the LOD values were used. The data prove that the removal of the nitrosamine precursors by the applied treatment processes seems to be much more efficient than for natural organic matter in general. Although removal of organic compounds in the two plants under investigation seems to be not extremely high (ca. 30% with respect to DOC and ca. 50% with respect to UV_{254}; however, it has to be taken into account that the absolute values for both parameters are rather low), the removal efficiencies for NDMA precursors (and probably also for NPYR precursors) are above 85%.

UTILITY E

From utility E samples from a groundwater treatment plant were received. In this plant (see Figure 7.8), the water is treated through a conventional filtration process (coagulation with ferric chloride, cationic polymer (polyamine blend), and 15 mg/L aluminum sulfate, flocculation, sedimentation, and filtration). Free chlorine is used as the primary disinfectant (applied after sedimentation), and chloramine is used in the distribution system.

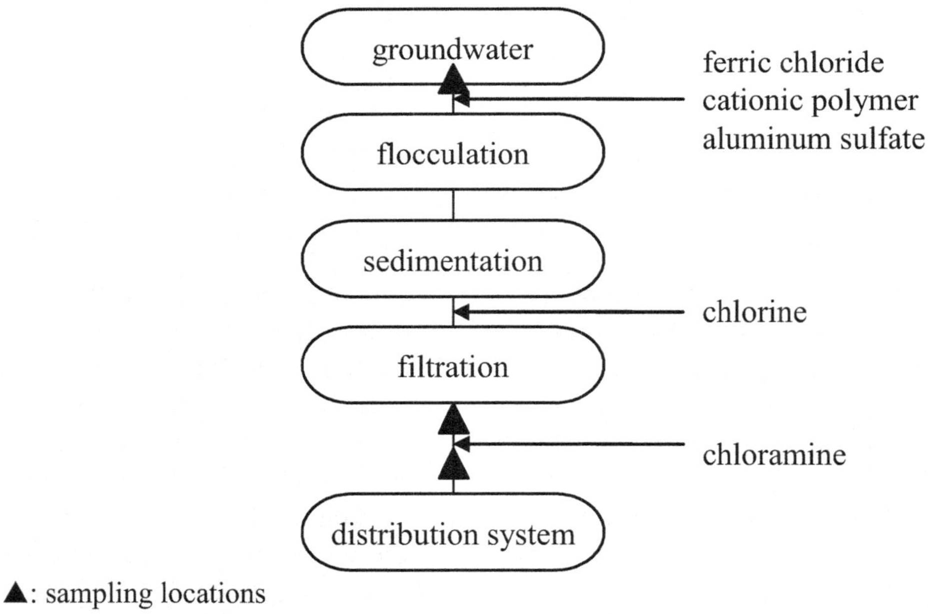

Figure 7.8 Treatment scheme of the groundwater treatment plant of utility E

Water quality data as well as nitrosamine concentrations and results of the NFP tests are given in Tables 7.8, 7.9, and 7.10 (sample D was only analyzed for its original nitrosamine content, but no NFP test was performed).

Table 7.8
Physical-chemical characteristics of water samples from utility E

Water sample (sampling date)	pH	Alkalinity (mg/L)	Ammonia (mg/L N)	Nitrate (mg/L N)	Nitrite (mg/L N)	TOC (mg/L)	color (C.U.)	Conductivity (μS/cm)	Turbidity (NTU)
Sample A raw water (09.02.2006)	7.73	147	0.17	0.06	0.010	1.1	12	460	0.31
Sample B settled raw water (09.02.2006)	7.78	146	<0.1	0.11	0.007	1.25	39	461	1.31
Sample C treated water (09.02.2006)	7.91	151	0.13	0.02	0.005	0.89	7	458	0.08
Sample D after disinfection (09.02.2006)	7.73	133	0.46	0.12	0.001	0.93	1	472	0.05

Source: Data provided by utility E

Table 7.9
Nitrosamine concentrations in original water samples from utility E

Water sample	NDMA (ng/L)	NEMA (ng/L)	NDEA (ng/L)	NDPA (ng/L)	NDBA (ng/L)	NPIP (ng/L)	NPYR (ng/L)	NMOR (ng/L)
Sample A raw water	4.9	<2	<2	<1	<1	<1	<1	<1
Sample B settled raw water	4.0	<2	<2	<1	<1	<1	<1	<1
Sample C treated water	<1	<2	<2	<1	<1	<1	<1	<1
Sample D after disinfection	1.4	<2	<2	<1	<1	<1	<1	<1

Table 7.10
NFP of water samples from utility E

Water sample	NDMA (ng/L)	NEMA (ng/L)	NDEA (ng/L)	NDPA (ng/L)	NDBA (ng/L)	NPIP (ng/L)	NPYR (ng/L)	NMOR (ng/L)
Sample A raw water	110	3.4	<2	<1	<1	<1	1.0	<1
Sample B settled raw water	13	<2	<2	<1	<1	<1	<1	<1
Sample C treated water	89	<2	<2	<1	<1	<1	1.6	<1

From the data in Table 7.9 it can be seen that NDMA could already be detected in the two original raw water samples of utility E (samples A and B). The origin for these NDMA findings in the raw water samples under investigation is unclear. Other nitrosamines than NDMA, however, were not detected in these waters. Interestingly, the NDMA levels in the disinfected water (sample D) are lower than in the raw waters, i.e. it seems that NDMA is removed during the applied treatment process and is formed again in small amounts during the disinfection step. However, the small differences may be attributable to variability in source levels. After chloramination under the conditions of the NFP test, NDMA as well as a small amount of NEMA could be detected in sample A. In the treated water (sample C) that was taken after filtration but before chloramination, the NDMA level formed during the NFP test has decreased and no NEMA is formed anymore, i.e. the treatment in the plant under investigation (flocculation, sedimentation and chlorination) leads to a significant removal of precursor compounds for nitrosamine formation. As for utility D, a comparison of the NFP data for the raw water and the treated water does not allow distinguishing between removal efficiencies of the single treatment steps but enable evaluation of the overall process. From the results of the laboratory-scale experiments, however, it can be concluded that chlorination is the decisive step for precursor removal whereas flocculation should be quite inefficient.

A comparison of removal efficiencies yields that TOC removal as well as removal of nitrosamine precursors is much lower for the plant of utility E than it was for the two plants of utility D. For the plant of utility E, removal rate for TOC as well as for NDMA precursors (measured via the NFP test) is only 19%, whereas it was ca. 30% for TOC and more than 85% for NDMA precursors at the two plants of utility D. However, a low TOC removal as it was found for utility E can be expected when the influent TOC is low.

UTILITY F

Utility F treats reservoir water (source water A) which is sometimes blended with river water (source water B). Typically, the river water has a higher color and turbidity compared to the reservoir water. During the pre-treatment process in the WTP, chlorine, ammonia, and lime are injected into the untreated water. After the injection of pre-treatment chemicals, the water is rapidly mixed for flocculation with aluminum sulfate (40 to 60 ppm depending on season). Residence time is 30 to 50 minutes depending on basin type. Flocculation is followed by sedimentation and filtration (filter effluent flume). After filtration, chlorine is added as primary disinfectant, followed by addition of lime, fluoride and orthophosphate (central flume). Finally, chlorine and ammonia are added for final disinfection by chloramine. Figure 7.9 gives a schematic overview of the treatment train of utility F and the locations of the four sampling points. Nitrosamine concentrations of the original water samples and results of the NFP tests are given in Tables 7.11 and 7.12.

The data in Table 7.11 indicate that no nitrosamines were present in the original water samples under investigation. After chloramination under the conditions of the NFP test, NDMA could be detected in all samples. Other nitrosamines than NDMA were not found. The NFP for the two source waters are of comparable order of magnitude. Compared to the raw water, there seems to be a slight increase in the filter effluent flume (treated water before disinfection). This slight increase in NDMA formation potential might be explained by some precursor release from the filters (made of sand, gravel and anthracite). In any case, these findings confirm the previous results that the conventional water treatment by flocculation, sedimentation, and filtration does not significantly reduce the amount of nitrosamine precursors. The NFP in the central flume (finished water before chloramine disinfection) however, is significantly lower than in the filter effluent and in the raw waters indicating that chlorine used as primary disinfectant reduces the NDMA precursor levels.

In conclusion, the results for utility F confirm the findings from the other utilities and the results from the laboratory-scale experiments indicating no impact on NFP by coagulation, flocculation and sedimentation but a significant reduction of nitrosamine precursor levels by chlorine oxidation.

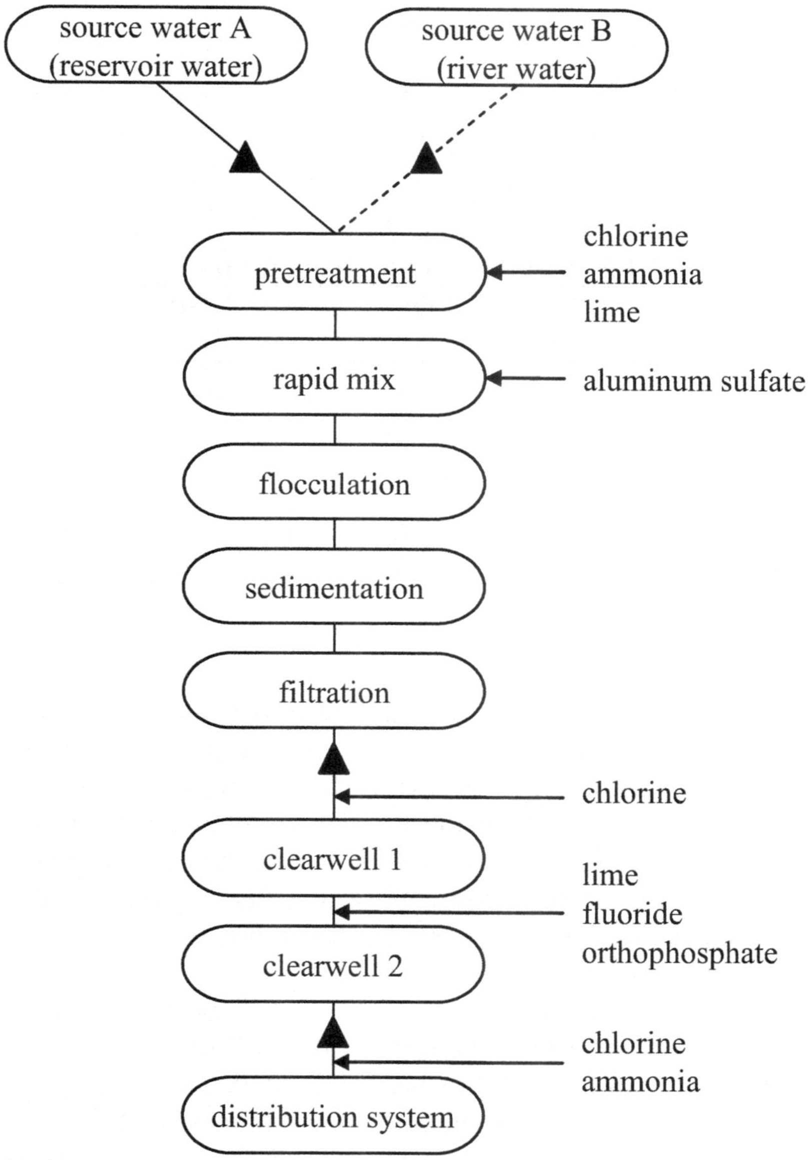

▲: sampling locations

Figure 7.9 Treatment scheme of the treatment plant of utility F

Table 7.11
Nitrosamine concentrations in original water samples from utility F

Water sample (sampling date)	NDMA (ng/L)	NEMA (ng/L)	NDEA (ng/L)	NDPA (ng/L)	NDBA (ng/L)	NPIP (ng/L)	NPYR (ng/L)	NMOR (ng/L)
Source water A reservoir water (07.02.2006)	<1	<2	<2	<1	<1	<1	<1	<1
Source water B river water (07.02.2006)	<1	<2	<2	<1	<1	<1	<1	<1
Filter effluent flume (07.02.2006)	<1	<2	<2	<1	<1	<1	<1	<1
Central flume (07.02.2006)	<1	<2	<2	<1	<1	<1	<1	<1

Table 7.12
NFP of water samples from utility F

Water sample	NDMA (ng/L)	NEMA (ng/L)	NDEA (ng/L)	NDPA (ng/L)	NDBA (ng/L)	NPIP (ng/L)	NPYR (ng/L)	NMOR (ng/L)
Source water A reservoir water	21	<2	<2	<1	<1	<1	<1	<1
Source water B river water	25	<2	<2	<1	<1	<1	<1	<1
Filter effluent flume	36	<2	<2	<1	<1	<1	<1	<1
Central flume	16	<2	<2	<1	<1	<1	<1	<1

UTILITY G

Utility G treats bank filtered water from shallow vertical and horizontal wells next to a river. The bank filtered water is pumped to one of two treatment plants. In both plants, the water is treated by aeration, ammonia addition (if necessary), lime softening, chlorine addition (between 2 and 4 mg/L total chlorine; 2.2 mg/L at the date of sampling), pH adjustment with carbon dioxide, and filtration through sand and gravel filters. Before the water is pumped into the distribution system, orthophosphate for corrosion control and fluoride are added (see Figure 7.10). Water quality data as well as nitrosamine concentrations of the original water samples and results of the NFP tests are given in Tables 7.13, 7.14, and 7.15.

As the data in Table 7.14 show, no nitrosamines were present in the two original water samples. After chloramination under the conditions of the NFP test, NDMA could be detected in both samples. Other nitrosamines than NDMA were not found. The NFP for the two samples are of comparable order, indicating that the applied treatment process (aeration and lime softening) does not significantly influence the precursor compounds. The already relatively low NFP values of the two samples provide an indication for the good water quality of the bank filtered water treated by utility G.

In conclusion, the results for utility G indicate that neither aeration nor lime softening result in a significant removal of precursors for nitrosamine formation during disinfection.

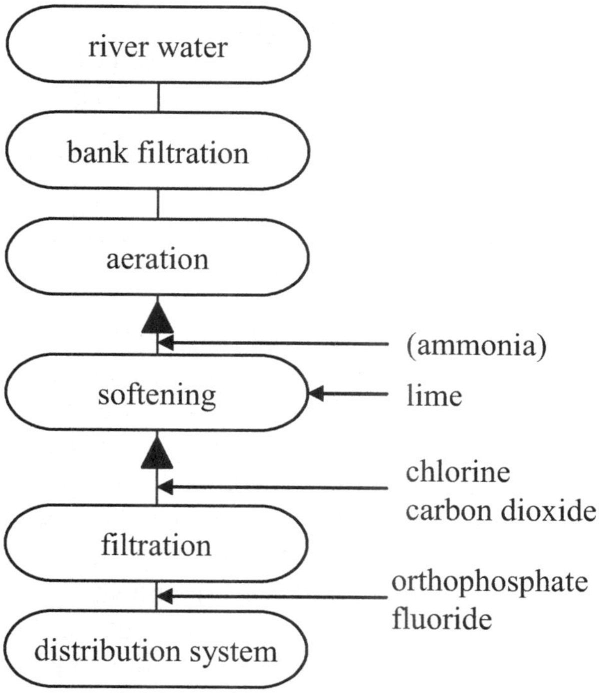

▲: sampling locations

Figure 7.10 Treatment scheme of the treatment plant of utility G

Table 7.13
Physical-chemical characteristics of water samples from utility G

Water sample (sampling date)	pH	Alkalinity (mg/L)	Total hardness (mg/L)	$CaCO_3$ hardness (mg/L)	Turbi-dity (NTU)	Conduc-tivity (μS/cm)	Nitrate (mg/L N)	Nitrite (μg/L N)	Ammonia (mg/L N)
raw influent before aeration (17.04.2006)	7.56	232	286	190	2.18	613	4.18	28.7	0.66
softener effluent (17.04.2006)	9.50	83	137	56	10.2	349	4.53	23.7	0.95

Source: Data provided by utility G

Table 7.14
Nitrosamine concentrations in original water samples from utility G

Water sample	NDMA (ng/L)	NEMA (ng/L)	NDEA (ng/L)	NDPA (ng/L)	NDBA (ng/L)	NPIP (ng/L)	NPYR (ng/L)	NMOR (ng/L)
raw influent	<1	<2	<2	<1	<1	<1	<1	<1
softener effluent	<1	<2	<2	<1	<1	<1	<1	<1

Table 7.15
NFP of water samples from utility G

Water sample	NDMA (ng/L)	NEMA (ng/L)	NDEA (ng/L)	NDPA (ng/L)	NDBA (ng/L)	NPIP (ng/L)	NPYR (ng/L)	NMOR (ng/L)
raw influent	16	<2	<2	<1	<1	<1	<1	<1
softener effluent	20	<2	<2	<1	<1	<1	<1	<1

UTILITY H

Utility H takes water from two surface water sources and – to a smaller extent – groundwater from a number of deep wells for water supply. Surface water sources are a river water and a reservoir water. The reservoir is filled exclusively by the river water and is used as a contingency in case of problems with river water quality or quantity. The treatment in the plant of utility H is conventional (see Figure 7.11 for a schematic presentation). The waters from different sources go directly into the plant aerators. After aeration, the pH of the water is raised by addition of lime and lowered again by addition of carbon dioxide. This treatment step increases hardness and alkalinity and guaranties optimum pH for the subsequent coagulation/flocculation process. Chlorine dioxide (0.8 mg/L) is used as pre-oxidant, and aluminum sulfate as coagulant. Before leaving the treatment plant, gaseous chlorine and aqueous ammonia solution are added stoichiometrically.

For utility H, a flow proportional source composite of about 90% river water and 10% water from a deep well as well as a treated water were investigated. The respective water quality parameters are given in Table 7.16. Tables 7.17 and 7.18 summarize nitrosamine concentrations in the original water samples and results of the NFP tests.

It can be seen from the data in Table 7.17 that neither the source water nor the treated water of utility H contained any of the nitrosamines under investigation. After chloramination, NDMA was found in both samples whereby the NDMA formation potential for the treated water was significantly lower than for the source water, indicating a good removal efficiency of the treatment process for nitrosamine precursors. The high removal efficiency of the treatment plant can also be seen from the decrease in TOC (Table 7.16). Based on the results of the laboratory-scale experiments it can be concluded that pre-oxidation by chlorine dioxide is the main treatment step that reduces the nitrosamine FP, whereas flocculation is only of minor relevance. As a result, the NDMA formation potential for the treated water of utility H is in the same range as for the treated waters of many other utilities under investigation.

Table 7.16
Physical-chemical characteristics of water samples from utility H

Water sample (sampling date)	pH	Alkalinity (mg/L)	Ammonia (mg/L N)	Nitrate (mg/L N)	Nitrite (mg/L N)	TOC (mg/L)
River water (21.03.2006)	6.35	8.5	0.05	0.88	<0.003	3.51
Reservoir water (21.03.2006)	6.62	40.5	0.05	0.79	<0.003	1.83

Source: Data provided by utility H

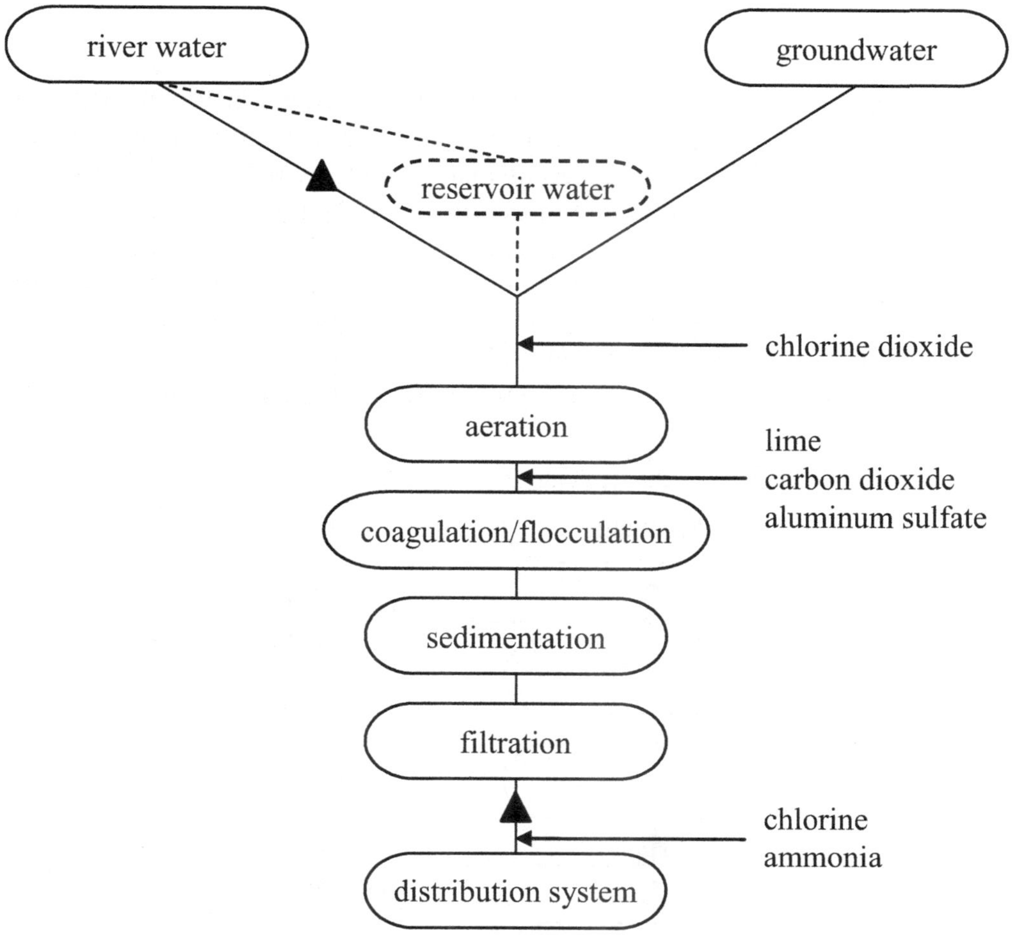

▲: sampling locations

Figure 7.11 Treatment scheme of the treatment plant of utility H

Table 7.17
Nitrosamine concentrations in original water samples from utility H

Water sample	NDMA (ng/L)	NEMA (ng/L)	NDEA (ng/L)	NDPA (ng/L)	NDBA (ng/L)	NPIP (ng/L)	NPYR (ng/L)	NMOR (ng/L)
River water	<1	<2	<2	<1	<1	<1	<1	<1
Reservoir water	<1	<2	<2	<1	<1	<1	<1	<1

Table 7.18
NFP of water samples from utility H

Water sample	NDMA (ng/L)	NEMA (ng/L)	NDEA (ng/L)	NDPA (ng/L)	NDBA (ng/L)	NPIP (ng/L)	NPYR (ng/L)	NMOR (ng/L)
River water	51	<2	<2	<1	<1	<1	<1	<1
Reservoir water	12	<2	<2	<1	<1	<1	<1	<1

Utility I

Utility I operates two treatment plants, each of them using a different surface water source. In WTP A, river water is treated by GAC filtration and direct filtration after addition of iron coagulants. After filtration, chlorine is added as disinfectant. In WTP B, a lake water is treated by iron based coagulation, flocculation and direct filtration. Again chlorine is added prior to distribution for disinfection purposes. The treatment trains of both plants of utility I are schematically presented in Figures 7.12 and 7.13. Water quality parameters for the samples under investigation are summarized in Table 7.19. Nitrosamine concentrations of the original water samples and results of the NFP tests are given in Tables 7.20 and 7.21. In addition to the samples from the two treatment plants, a distributed water from a system using chloramination (neither WTP A nor WTP B) was analyzed.

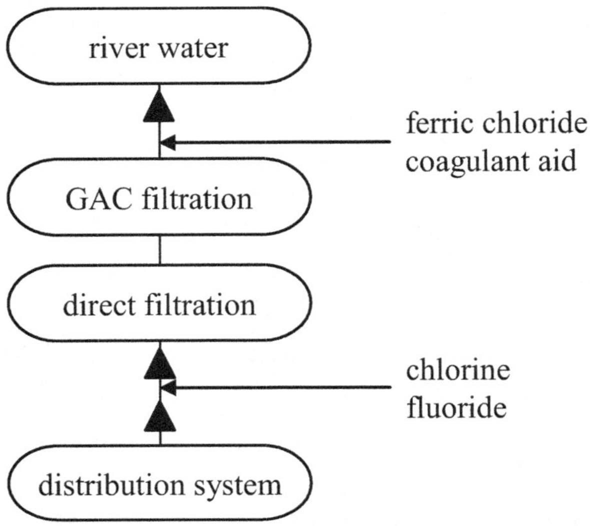

▲: sampling locations

Figure 7.12 Treatment scheme of treatment plant A of utility I

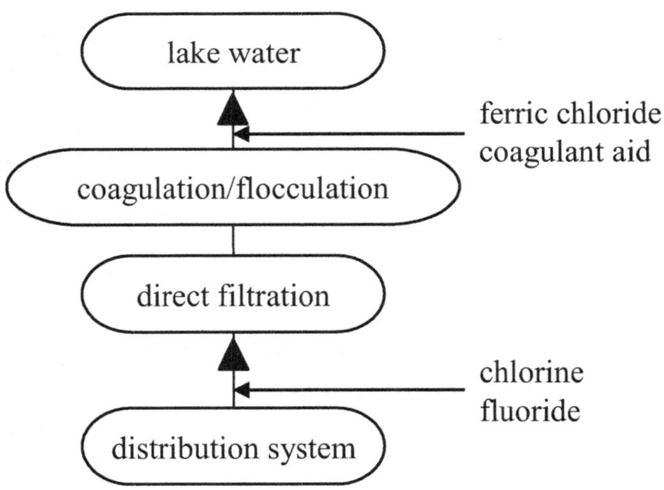

▲: sampling locations

Figure 7.13 Treatment scheme of treatment plant B of utility I

Table 7.19
Physical-chemical characteristics of water samples from utility I

Water sample (sampling date)	pH	Alkalinity (mg/L)	Ammonia (mg/L N)	Nitrate (mg/L N)	Nitrite (mg/L N)	TOC (mg/L)
WTP A raw water (04.05.2006)	7.3	34.5	0.01	0.17	0.002	3.5
WTP A treated water (04.05.2006)	7.2	33.0	<0.01	0.19	0.001	2.1
WTP A finished water (04.05.2006)	7.4	36.0	0.01	0.18	0.001	1.6
WTP B raw water (04.05.2006)	7.0	41.5	0.01	0.18	0.001	3.3
WTP B treated water (04.05.2006)	8.0	50.5	<0.01	0.18	0.001	2.7
WTP distributed water (04.05.2006)	7.6	40.0	0.3	0.16	0.001	2.8

Source: Data provided by utility I

Table 7.20
Nitrosamine concentrations in original water samples from utility I

Water sample	NDMA (ng/L)	NEMA (ng/L)	NDEA (ng/L)	NDPA (ng/L)	NDBA (ng/L)	NPIP (ng/L)	NPYR (ng/L)	NMOR (ng/L)
WTP A raw water	<1	<2	<2	<1	<1	<1	<1	<1
WTP A treated water	<1	<2	<2	<1	<1	<1	<1	<1
WTP A finished water	<1	<2	<2	<1	<1	<1	<1	<1
WTP B raw water	<1	<2	<2	<1	<1	<1	<1	<1
WTP B treated water	<1	<2	<2	<1	<1	<1	<1	<1
WTP distributed water	1.9	<2	<2	<1	<1	<1	<1	<1

Nitrosamine analyses in the original water samples (Table 7.20) indicated that only in the distributed water from a treatment plant using chloramination as final disinfection step small amounts of NDMA could be detected. All other water samples under investigation were free of NDMA and other nitrosamines than NDMA were not present.

Table 7.21
NFP of water samples from utility I

Water sample	NDMA (ng/L)	NEMA (ng/L)	NDEA (ng/L)	NDPA (ng/L)	NDBA (ng/L)	NPIP (ng/L)	NPYR (ng/L)	NMOR (ng/L)
WTP A raw water	19	<2	<2	<1	<1	<1	<1	<1
WTP A treated water	24	<2	<2	<1	<1	<1	<1	<1
WTP A finished water	23	<2	<2	<1	<1	<1	<1	<1
WTP B raw water	8.4	<2	<2	<1	<1	<1	<1	<1
WTP B treated water	88	<2	<2	<1	<1	<1	<1	<1
WTP distributed water	8.7	<2	<2	<1	<1	<1	<1	<1

After chloramination under the conditions of the NFP test, NDMA was found in all samples. In treatment plant A, the NDMA levels were rather constant from the raw water to the finished water, indicating that the applied treatment by flocculation and filtration does not result in a significant elimination of precursor compounds. Even GAC filtration does not lead to a reduction in NFP, and thus no removal of NDMA precursors can be stated. For treatment plant B, the concentrations of NDMA formed in the NFP test increase dramatically from raw water to treated water but decrease again in the finished water. A possible explanation for the observed effect might be that the coagulant aid used in this plant releases organic compounds that act as NDMA precursors. In literature it is well described that polyelectrolytes used in water treatment as primary coagulants or as floc builders might be important source of NDMA precursors during drinking water treatment. However, as no information on the chemical nature of the coagulant aid used in WTP B is available, it is not possible to draw a final conclusion from these results. In any case, it seems as if the chlorination step applied in WTP B reduces the NDMA formation potential again. A similar behavior was already observed for other utilities.

CHAPTER 8
STRATEGIES FOR MINIMIZING NITROSAMINE FORMATION DURING DISINFECTION OF DRINKING WATER AND FUTURE RESEARCH NEEDS

STRATEGIES FOR MINIMIZING NITROSAMINE FORMATION DURING DISINFECTION OF DRINKING WATER

Nitrosamine formation is especially crucial for water suppliers using chloramination for disinfection of their finished water. Disinfection, however, is essential for safeguarding the microbiological quality of the drinking water at consumers' tap. Thus, strategies are needed that enable minimizing nitrosamine formation during chloramination without compromising disinfection efficiency. Looking at the results of this research project it becomes clear that NDMA is by far the most relevant nitrosamine formed during chloramination and thus minimization strategies should mainly focus on NDMA formation. In laboratory-scale experiments as well as in several full-scale tests, the levels of NDEA, NEMA, NDPA, NDBA, NPIP, NPYR, and NMOR formed during chloramination were one or two orders of magnitude lower than the concentrations of NDMA and in most cases no other nitrosamines than NDMA could be found.

From the outcome of the literature search as well as from the results of the laboratory-scale experiments and the full-scale tests at several water utilities, several options for minimizing nitrosamine formation became visible. The first option takes into account the raw water source. Most of the precursors for nitrosamine formation during chloramination can be related to anthropogenic sources and thus water utilities using surface waters that are characterized by high wastewater loads are more prone to nitrosamine formation than those using non-impacted water sources as raw water for drinking water production. Water utilities that have no alternative to using wastewater-impacted surface water might utilize treatment processes as slow sand filtration, artificial groundwater recharge or riverbank filtration to reduce levels of nitrosamine precursors. Besides the removal of precursors, these treatment steps would also reduce concentration levels of nitrosamines already present in the surface waters.

Optimizing operational parameters of the chloramination process is another possibility for minimizing nitrosamines levels in the disinfected waters. It is obvious that decreasing the disinfectant dose leads to a reduction of nitrosamine levels. Lowering the pH is another option, as nitrosamine formation during chloramination is more efficient at pH values of 8 and above than at a pH of around 7. When changing operational parameters, however, the microbiological safety of the disinfected water must always be taken into account and thus the possibilities of a water supplier are most often limited. One interesting option for minimizing nitrosamine formation during chloramination might be the production of chloramines in a manner that minimizes the concentration of dichloramine, e.g. by addition of free chlorine before ammonia. In literature it is reported that especially dichloramine leads to the formation of elevated levels of nitrosamines and thus by avoiding the formation of this species might lead to a reduction in nitrosamine formation.

In most cases, the use of chlorine (or hypochlorite) as disinfectant instead of chloramine will also lead to a significantly lower nitrosamine formation. However, it has to be taken into account that chlorine produces other by-products of toxicological relevance and that chlorine has disadvantages with regard to stability in the distribution system.

Table 8.1
Removal efficiencies of single treatment steps for nitrosamine precursors

Treatment step	Removal efficiency (%)	
	Laboratory-scale experiment	Full-scale test
Slow sand filtration, artificial groundwater recharge, riverbank filtration	10 – 30	80 – 93
Aeration	n.i.	< 10
Lime softening	n.i.	< 10
Flocculation (iron and aluminum salts)	< 10	< 10
Oxidation		
Ozone	30 – 92	ca. 90
Chlorine dioxide	48 – 94	n.i.
Chlorine/Hypochlorite	57 – 94	19 – 87
Iron(VI)	40 – 84	n.i.
Activated carbon adsorption	10 – > 90 (depending on AC dose)	< 10

n.i.: not investigated

A promising strategy for minimizing nitrosamine formation during chloramination is the removal of precursor compounds prior to disinfection. Table 8.1 summarizes the efficiencies of several treatment steps for removal of NDMA precursors. During the laboratory-scale experiments it was found that removal efficiencies for other nitrosamine precursors are comparable to those for NDMA precursors and thus the results in Table 8.1 can be generalized.

As already mentioned above, microbiological degradation during slow sand filtration, artificial groundwater recharge, or riverbank filtration is an interesting option for removal of nitrosamine precursors. Results from full-scale tests at water utilities indicate an even higher removal rate than it was found during laboratory-scale experiments.

Full-scale treatment data indicate that aeration or lime softening treatment steps have no impact on the nitrosamine formation potential and thus can not be used for minimizing nitrosamine formation during chloramination.

Flocculation also turned out to be inefficient in removal of nitrosamine precursors. In contrast, it is reported in literature that polyelectrolytes used in water treatment as primary coagulants or as floc builders, are important sources of NDMA precursors. It was shown that especially the use of polyDADMAC and epi-DMA in the coagulation/flocculation/sedimentation process causes significant NDMA formation during chlorination and chloramination. Thus, the use of these polymer materials should be avoided or at least their doses should be optimized if the treated water is disinfected with chlorine or chloramine.

A much better option for minimizing nitrosamine formation is the inclusion of a pre-ozonation or pre-chlorination step in the treatment process. Laboratory-scale experiments as well as the full-scale tests at several water utilities clearly proved that oxidation is an efficient tool for removal of nitrosamine precursors. Observed reduction rates for NDMA formation after pre-ozonation as well as after pre-chlorination amounted to more than 90%. Both treatment steps, however, produce other by-products than nitrosamines and thus the whole treatment process and all unwanted by-products must always be taken into account.

From the results of the laboratory-scale adsorption experiments it can be concluded that adsorption onto activated carbon might be a promising option for the – at least partial – removal of precursor compounds for nitrosamine formation during chloramination. These results, however, could not be confirmed by the full-scale test at one water utility and thus further investigations are necessary to assess the efficiency of full-scale GAC filtration for precursor removal.

When looking at the possible treatment options for removal of nitrosamine precursors, it should always been taken into account that these processes not only efficiently minimize nitrosamine formation during chloramination but remove a lot of other unwanted compounds from the raw water.

Removal of nitrosamines formed during chloramination of drinking waters can not be regarded as a primary option. Chloramination in most cases is the final treatment step in a waterworks and thus an additional post-treatment is not feasible. Furthermore, the options for nitrosamine removal are limited as NDMA and other nitrosamines are not easily removed by conventional treatment technologies like flocculation, ozonation or GAC filtration. UV irradiation is effective in nitrosamine removal but required UV doses are extremely high and the main applications of this promising technology seem to be the remediation of contaminated sites rather than removal of trace nitrosamines from drinking water.

In conclusion, the main strategies for minimizing nitrosamine formation during disinfection of drinking waters seem to be the use of raw waters not impacted by industrial or municipal wastewaters, the optimization of operational parameters of the disinfection process, the avoidance of polymeric materials during the flocculation process, or the efficient removal of nitrosamine precursors by processes as slow sand filtration, artificial groundwater recharge, riverbank filtration, or by pre-chlorination or pre-ozonation.

FUTURE RESEARCH NEEDS

Based on the outcomes of this study, some knowledge gaps and needs for further research related to the formation of nitrosamines during chloramination and the identification of possible strategies for minimizing nitrosamine formation can be identified. This research should focus on the formation of NDMA as according to the results of this project other nitrosamines are only of minor relevance.

Although a lot of research has been done in the last years on possible mechanisms and pathways for the formation of nitrosamines during chloramination, there are still some missing pieces. A complete understanding of the pathways and the key compounds and intermediates involved in the formation process, however, will enable to identify possible strategies for minimizing nitrosamine formation.

During the laboratory-scale experiments described in this report, some key elements in the chemical structure of possible precursor compounds for nitrosamine formation have been identified. Furthermore, it could be demonstrated that NDMA formation during chloramination most likely can be attributed to one single compound or a group of very similar compounds. The identification of this major precursor and its major sources, however, is still lacking. Elimination of this precursor before disinfection would be the optimum strategy to avoid nitrosamine formation.

During the full-scale tests, especially at U.S. waterworks, trace-levels of nitrosamines, especially NDMA have been detected already in the source waters. Future research should try to identify the sources for these nitrosamines. If other sources for nitrosamines than chloramination are identified, then more research on treatment options for removal of nitrosamines should follow.

The full-scale test at one specific plant showed that chemicals used as coagulant might contribute to a significant NDMA formation during chloramination. Further research is needed to understand the underlying processes and to identify products which lead to elevated NDMA formation.

Nitrosamine formation during other treatment steps where oxidants are applied should be studied in more detail. Nitrosamine formation during chlorination in the absence of ammonia still is questionable and other oxidation processes might also lead to the formation of nitrosamines.

Although nitrosamines are of major relevance for drinking water utilities because of their high toxicological impact, other nitrogen-containing disinfection should not be completely neglected. Looking at the proposed pathways for nitrosamine formation during chloramination, other intermediates and by-products can be identified which have not been assessed for their relevance yet.

REFERENCES

Abalos, M., J.M. Bayona, and F. Ventura. 1999. Development of a solid-phase microextraction GC-NPD procedure for the determination of free volatile amines in wastewater and sewage-polluted waters. *Analytical Chemistry*, 71: 3531-3537.

Abia, L., X.L. Armesto, M. Canle, M.V. Garcia, and J.A. Santaballa. 1998. Oxidation of aliphatic amines by aqueous chlorine. *Tetrahedron*, 54: 521-530.

Abidi, S.L. 1982. Detection of diethylnitrosamine in nitrite-rich water following treatment with rhodamine flow tracers. *Water Research*, 16: 199-204.

Acero, J.L., and U. von Gunten. 2001. Characterization of oxidation processes: Ozonation and the AOP O_3/H_2O_2. *Jour. AWWA*, 93(10): 90-100.

Andrews, S.A. and V.Y. Taguchi. 2000. NDMA – Canadian issues. Proceedings 2000 WGTC, Salt Lake City, Utah.

Angeles, R.M., L.K. Keefer, P.R. Roller, and S.J. Uhm. 1978. Chemical models for possible nitrosamine artifact formation in environmental analysis. Pages 109-115 in E. A. Walker and L. Griciute, eds. Environmental aspects of N-nitroso compounds. Lyon: IARC Press.

Archer, M.C., H.S. Yang, and J.D. Okun. 1978. Acceleration of nitrosamine formation at pH 3.5 by microorganisms. Pages 239-246 in E. A. Walker and L. Griciute, eds. Environmental aspects of N-nitroso compounds. Lyon: IARC Press.

Arienzo, M., J. Gan, F. Ernst, S. Qin, S. Bondarenko, and D.L. Sedlak. 2006. Loss pathways of N-nitrosodimethylamine (NDMA) in turfgrass soils. *Jour. of Environmental Quality*, 35: 295-292.

ATSDR. 1989. Agency for Toxic Substances and Disease Registry: Toxicological profile for N-nitrosodipropylamine.

Ayanaba, A., and M. Alexander. 1974. Transformations of methylamines and formation of a hazardous product, dimethylnitrosamine, in samples of treated sewage and lake water. *Jour. of Environmental Quality*, 3: 83-89.

Ayanaba, A., W. Verstraeten, and M. Alexander. 1973. Formation of dimethylnitrosamine, a carcinogen and mutagen, in soils treated with nitrogen-compounds. *Soil Science Society of America Jour.*, 37: 565-568.

Bader, H., and J. Hoigné. 1981. Determination of ozone in water by the indigo method. *Water Research*, 15: 449-456.

Bamforth, C.W. and P.J. Large. 1977. Solubilization, partial purification and properties of N-methylglutamate dehydrogenase from pseudomonas aminovorans. *Biochemical Jour.*, 161: 357-370.

Banerjee, S., E.J. Pack, H. Sikka, and C.M. Kelly. 1984. Kinetics of oxidation of methylhydrazines in water. Factors controlling the formation of 1,1-dimethylnitrosamine. *Chemosphere*, 13: 549-559.

Barrett, S., C. Hwang, Y. Guo, S.A. Andrews, and R.L. Valentine. 2003. Occurrence of NDMA in drinking water: A North American survey, 2001 – 2002. Proceedings 2003 AWWA Annual Conference, Anaheim, Calif.

Bergstrom, P.D., D.W. Grant, and S.M. Morrison. 1972. Nitrosation in feedlot manure. *Environmental Letters*, 3: 151-157.

Berman, T., and D.A. Bronk. 2003. Dissolved organic nitrogen: a dynamic participant in aquatic ecosystems. *Aquatic Microbial Ecology*, 31: 279-305.

Bircher, K.G., S.R. Carter, W. Lem, D. Dolan, and M. Crawford. 1999. Treatment of NDMA in drinking water by a Rayox UV tower. Proceedings Perchlorate Conference, East Valley Municipal Water District, Ontario, California.

Bittersohl, G., and H. Heberer. 1978. Zur Toxizität von aliphatischen Aminen. *Zeitschrift für die gesamte Hygiene und ihre Grenzgebiete*, 24: 529-534.

Börnick, H., and T.C. Schmidt. 2006. Amines. In T. Reemtsma and M. Jekel: Organic pollutants in the water cycle. Wiley-VCH, Weinheim. pp. 181-209

Boethling, R.S., and M. Alexander. 1979. Microbial degradation of organic compounds at trace-levels. *Environmental Science & Technology*, 13: 989-991.

Bolto, B., D. Dixon, R. Eldridge, and S. King. 2002. Removal of THM precursors by coagulation or ion exchange. *Water Research*, 36: 5066-5073.

Bolton, J.R., K.G. Bircher, W. Tumas, and C.A. Tolman. 2001. Figures-of-merit for the technical development and application of advanced oxidation processes. *Pure & Applied Chemistry*, 76: 627-637.

Bolton, J.R., S.R. Fitzsimmons, M. Patel, W. Lem, and J. Gigliotti. 2002. Experimental determination of the fluence (UV Dose) required in the UV photodegradation of NDMA in a treated waste water. Water Quality Technology Conference. American Water Works Association.

Boulton, C.A., and P.J. Large. 1977. Synthesis of certain assimilatory and dissimilatory enzymes during bacterial adaptation to growth on trimethylamine. *Jour. of General Microbiology*, 101: 151-156.

Boyland, E., E. Nice, and K. Williams. 1971. The catalysis of nitrosation by thiocyanate from Saliva. *Food and Cosmetics Toxicology*, 9: 639-643.

Brooks, J.B., W.B. Cherry, L. Thacker, and C.C. Alley. 1972. Analysis by gas chromatography of amines and nitrosamines produced in vivo and in vitro by Proteus vulgaris. *Journal of Infectious Diseases*, 126: 143-153.

Brunnemann, K.D., S.S. Hecht, and D. Hoffmann. 1982. N-Nitrosamines – Environmental occurrence, invivo formation and metabolism. *Jour. of Toxicology-Clinical Toxicology*, 19: 661-688.

Buxton, G.V., and C.L. Greenstock. 1988. Critical review of rate constants for reactions of hydrated electrons, hydrogen atoms and hydroxyl radicals in aqueous solution. *Jour. of Physical and Chemical Reference Data*, 17: 513-886.

Calamari, D., R. da Gasso, S. Galassi, A. Provini, and M. Vighi. 1980. Biodegradation and toxicity of selected amines on aquatic organisms. *Chemosphere*, 9: 762.

Calgon Carbon Corporation. 2000. Full-scale ISEPI groundwater treatment plant – Project summary information.

Castegnaro, M., I. Brouet, J. Michelon, G. Lunn, and E.B. Sansone. 1986. Oxidative destruction of hydrazines produces N-nitrosamines and other mutagenic species. *American Industrial Hygiene Association Jour.*, 47: 360-364.

Cattaneo, P., C. Cantoni, and C. Bersani. 1981. Degradation of N-nitrosamines by bacteria. *Industrie Alimentari*, 20: 376-378.

Challis, B.C., A. Edwards, R.R. Hunma, S.A. Kyrtopoulos, and J.R. Outram. 1978. Rapid formation of N-nitrosamines from nitrogen oxides under neutral and alkaline conditions. Pages 127-142 in E. A. Walker and L. Griciute, eds. Environmental aspects of N-nitroso compounds. Lyon: IARC Press.

Charrois, J.W.A., M.W. Arend, K.L. Froese, S.E. Hrudey. 2004. Detecting N-nitrosamines in drinking water at nanogram per liter levels using ammonia positive chemical ionization. *Environmental Science and Technology*, 38:4835-4841.

Chen Z., and R.L. Valentine. 2005. Studies on the formation of N-nitrosodimethylamine (NDMA) from natural organic matter. Abstracts of papers of the American Chemical Society: U847, 132-ENVR Part 1.

Child, P., G. Kaa, D. Benitz, P. Fowlie, and R. Hong-You. 1996. Reaction between chlorine and dimethylamine containing polyelectrolyte leading to the formation of N-nitroso-dimethylamine. http://www.pollutechgroup.com/papers/p11.thm.

Choi, J., S. Duirk, and R.L. Valentine. 2002. Mechanistic studies of N-nitrosodimethylamine (NDMA) formation in chlorinated drinking water. *Jour. Environmental Monitoring*, 4: 249-252.

Choi, J., and R.L. Valentine. 2002a. A kinetic model of N-nitrosodimethylamine (NDMA) formation during water chlorination/chloramination. *Water Science and Technology*, 46: 65-71.

Choi, J., and R.L. Valentine. 2002b. Formation of N-nitrosodimethylamine (NDMA) from reaction of monochloramine: a new disinfection by-product. *Water Research*, 36: 817-824.

Choi, J., and R.L. Valentine. 2003. N-nitrosodimethylamine formation by free-chlorine-enhanced nitrosation of dimethylamine. *Environmental Science & Technology*, 37: 4871-4876.

Choudhary, G., and H. Hansen. 1998. Human health perspective on environmental exposure to hydrazines: A review. *Chemosphere*, 37: 801-843.

Cohen, J.B., and J.D. Bachman. 1978. Measurement of environmental nitrosamines. Pages 257-372 in E. A. Walker, ed. Environmental aspects of N-nitrosamines. IARC.

Colby, J., and L.J. Zatman. 1971. The purification and properties of a bacterial trimethylamine dehydrogenase. *Biochemical Jour.*, 121: 9-10.

Colby, J., and L.J. Zatman. 1973. Trimethylamine metabolism in obligate and facultative methylotrophs. *Biochemical Jour.*, 132: 101-112.

Cromwell, B.T. and M. Richardson. 1966. Studies on the biogenesis of some simple amines and quaternary ammonium compounds in higher plants. Trimethylamine in Chenopodium vulvuria L. *Phytochemistry*, 5: 735-746.

Davis, M.K., S.E. Barrett, C.J. Hwang, Y. Guo, S. Liang. 2000. N-Nitrosodimethylamine (NDMA) in surface water. Proceedings of American Water Works Association Water Quality Technology Conference 2000.

Dean-Raymond, D. and M. Alexander. 1976. Plant uptake and leaching of dimethylnitrosamine. *Nature*, 262: 394-396.

DHS. 2002. California Department of Health Services: Studies on the occurrence of NDMA in drinking water. http://www.dhs.ca.gov/ps/ddwem/chemicals/NDMA/studies.htm.

DHS. 2005. A brief history of NDMA findings in drinking water supplies and related activities. Available at http://www.dhs.ca.gov/ps/ddwem/chemicals/NDMA/history.htm. CDHS, Sacramento.

DHS. 2006. California's experience with NDMA in drinking water. http://www.dhs.ca.gov/ps/ddwem/chemicals/NDMA/NDMAindex.htm. CDHS, Sacramento.

DIN. 1994. German Standard Method DIN 32645: Chemical analysis; Decision limit, detection limit and determination limit. Beuth Publishing, Berlin, Germany.

Diyamandoglu, V. 1994. Nitrate and chloride formation in chloramination. *Water Science and Technology*, 30: 101-110.

Diyamandoglu, V., and R.E. Selleck. 1992. Reactions and products of chloramination. *Environmental Science & Technology*, 26: 808-814.

Dominguez, G. 2001. System uses UV light to destroy groundwater contamination. Center Views of the Air Force Center for Environmental Excellence 7(4): 1-2.

DVGW. 1998: Technical Rules, Working Sheet W218: Flockung in der Wasseraufbereitung, Teil 2: Flockungsverfahren. DVGW-Regelwerk.

Duré, G., L. Weil, and K.-E. Quentin. 1978. Modellversuche zum Verhalten von N-Nitrosaminen. Vom Wasser 50: 21-43.

Eady, R.R., T.R. Jarman, and P.J. Large. 1971. Microbial oxidation of amines. Partial purification of a mixed-function secondary-amine oxidase system from Pseudomonas aminovorans that contains an enzymically active cytochrome-P-420-type haemoprotein. *Biochemical Jour.*, 125: 449-459.

Eady, R.R., and P.J. Large. 1968. Purification and properties of an amine dehydrogenase from Pseudomonas AM1 and its role in growth on methylamine. *Biochemical Jour.*, 106: 245-255.

Eady, R.R., and P.J. Large. 1969. Bacterial oxidation of dimethylamine, a new mono-oxygenase reaction. *Biochemical Jour.*, 111: 37-38.

Eady, R.R., and P.J. Large. 1971. Microbial oxidation of amines. Spectral and kinetic properties of the primary amine dehydrogenase of Pseudomonas AM1. *Biochemical Jour.*, 123: 757-771.

Eisenbrand, G., B. Spiegelhalder, J. Kann, R. Klein, and R. Preussmann. 1979. Carcinogenic N-nitrosodimethylamine as a contamination in drugs containing 4-dimethyl-2,3-di-methyl-1-phenyl-3-pyrazolin-5-on (amidopyrine, aminophenazone). *Arzneimittel Forschung*, 19: 867-868.

Elmghari-Tabib, M., A. Laplanche, F. Venien, and G. Martin. 1982. Ozonation of amines in aqueous solutions. *Water Research*, 16: 223-229.

EPA. 1997. N-Nitrosodimethylamine CASNR 62-75-9, Integrated risk information service (IRIS) substance file. http://www.epa.gov/iris/subst/0045.htm.

Falbe, J., and M. Regitz (Eds.) 1995. Römpp Chemie Lexikon. Stuttgart: Thieme.

Fan, T.-Y., U. Goff, L. Song, D.H. Fine, P. Arsenault, and K. Bieman. 1977. N-Nitrosodiethanolamine in cosmetics, lotions and shampoos. *Food and Cosmetics Toxicology*, 15: 423-430.

Fan, T.-Y., and S. Tannenbaum. 1973. Factors influencing the rate of formation of nitrosomorpholine from morpholine and nitrite: Acceleration by thiocyanate and other amines. *Jour. Agricultural Food Chemistry*, 21: 237-240.

Fiddler, W., J.W. Pensabene, R.C. Doerr, and C.J. Dooley. 1977. Presence of dimethyl-nitrosamines and diethyl-nitrosamines in deionized water. *Food and Cosmetics Toxicology*, 15: 441-443.

Fishbein, L. 1979. Overview of some aspects of occurrence, formation and analysis of nitrosamines. *The Science of the Total Environment*, 13: 157-188.

Fleming, E.C., J.C. Pennington, B.G. Wachob, R.A. Howe, and D.O. Hill. 1996. Removal of N-nitrosodimethylamine from waters using physical-chemical techniques. *Jour. of Hazardous Materials*, 51: 151-164.

Freundlich, H. 1906. Über die Adsorption in Lösungen. *Zeitung für Physikalische Chemie*, 57:385-470.

Gan J., S. Bondarenko, F. Ernst, W. Yang, S.B. Ries, D.L. Sedlak. 2006. Leaching of N-nitrosodimethylamine (NDMA) in turfgrass soils during wastewater irrigation. *Jour. of Environmental Quality*, 35: 277-284.

Gerecke, A.C., and D.L. Sedlak. 2003. Precursors of N-nitrosodimethylamine in natural waters. *Environmental Science & Technology*, 37: 1331-1336.

Godoy, H.M., M.I.D. Gomez, and J.A. Castro. 1984. Metabolism and activation of 1,1-dimethylhydrazine and methylhydrazine, two products of N-nitrosodimethylamine reductive biotransformation. *IARC Scientific Publications*, 57: 479-484.

Golovnya, R.V., I.L. Zhuravleva, M.B. Terenina, N.L. Vlasenko, and A.P. Il'nitsky. 1982. Amines in soil as possible precursors of N-nitroso compounds. *IARC Scientific Publications*, 41: 327-335.

Gough, T.A., K.S. Webb, and M.F. McPhail. 1977. Volatile nitrosamines from ion exchange resins. *Food and Cosmetics Toxicology*, 15: 437-440.

Greene, S., M. Alexander, and D. Leggett. 1981. Formation of N-nitrosodimethylamine during treatment of municipal wastewater by simulated land application. *Jour. of Environmental Quality*, 10: 416-421.

Grilli, A., and G. Prodi. 1975. Identification of dimethylnitrosamine metabolites in vitro. *Gann Monograph on Cancer Research*, 66: 473-480.

Gui, L., R. Gillham, and M. Odziemkowski. 2000. Reduction of N-nitrosodimethylamine with granular iron and nickel-enhanced iron. 1. Pathways and kinetics. *Environmental Science & Technology*, 34: 3489-3494.

Gunnison, D., M. Zappi, C. Teeter, J. Pennington, and R. Bajpai. 2000. Attenuation mechanisms of N-nitrosodimethylamine at an operating intercept and treat groundwater remediation system. *Jour. of Hazardous Materials*, B73: 179-197.

Haag, W.R., and C.C.D. Yao. 1992. Rate constants for reaction of hydroxyl radicals with several drinking water contaminants. *Environmental Science & Technology*, 26: 1005-1013.

Haase, L.W. 1936. Chemisches von der Chlorung. Vom Wasser 11: 116-136.

Hampton, D., and L.J. Zatman. 1973. The metabolism of tetramethylammonium chloride by bacterium 5H2. *Biochemical Society Transactions*, 1: 667-668.

Harada, K. 1980. Microbial-degradation of nitrosamines. 2. Effect of the conditions of growth and enzymic reaction on the nitrosamine breakdown. *Bulletin of the Japanese Society of Scientific Fisheries*, 46: 723-726.

Harada, K., and K. Yamada. 1979. Microbial degradation of nitrosamines 1. Inducible breakdown of nitrosamines. *Bulletin of the Japanese Society of Scientific Fisheries*, 45: 925-928.

Havery, D.C., and H.J. Chou. 1994. Nitrosamines in sunscreens and cosmetic products – Occurrence, formation, and trends. *ACS Symposium Series*, 553: 20-33.

Hawksworth, G.M., and M.J. Hill. 1971. Bacteria and the N-nitrosation of secondary amines. *British Jour. for Cancer*, 25: 520-526.

Heur, Y. –H., A.J. Streeter, R.W. Nims, and L.K. Keefer. 1989. The Fenton degradation as a nonenzymatic model for microsomal denitrosation of N-nitrosodimethylamine. *Chemical research in toxicology*, 2: 247–253.

Ho, T.-F.L., J.R. Bolton, and E. Lipczynska-Kochany. 1996. Quantum yields for the photodegradation of pollutants in dilute aqueous solution: Phenol, 4-chlorophenol and N-nitrosodimethylamine. *Jour. of Advanced Oxidation Technology*, 1: 170-178.

Hoigné, J., and H. Bader. 1983. Rate constants of reactions of ozone with organic and inorganic compounds in water – II: Dissociating organic compounds. *Water Research*, 17: 185-194.

Hoigné, J., and H. Bader. 1994. Characterization of water quality criteria for ozonation processes. Part 2: Lifetime of added ozone. *Ozone Science & Engineering*, 16: 121-134.

Hoppe, C., T. Jennings, J.E. Drewes. 2006. Attenuation of organic micropollutants during riverbank filtration – Results from field studies and laboratory experiments. Proceedings NGWA – 5th International Conference on Pharmaceuticals and Endocrine Disrupting Chemicals in Water. March 13-15, Costa Mesa, Californien.

Howard, P.H., R.S. Boethling, W.F. Jarvis, W.M. Meylan, and E.M. Michalenko (Eds.) 1991. Handbook of environmental degradation rates. Chelsea, MI: Lewis Publishers Inc.

Huber, M.M., S. Canonica, G.-Y. Park, and U. von Gunten. 2003. Oxidation of pharmaceuticals during ozonation and advanced oxidation processes. *Environmental Science & Technology*, 37: 1016-1024.

Huber, S.A., and F.H. Frimmel. 1992a. A liquid chromatographic system with multi-detection for the direct analysis of hydrophilic organic compounds in natural waters. *Fresenius J. Anal. Chem.*, 342: 198-200.

Huber, S.A., and F.H. Frimmel. 1992b. A new method for the characterization of organic carbon in aquatic systems. *J. Environ. Anal. Chem.*, 49: 49-57.

Huie, R.E., and P. Neta. 1986. Rate Constants for one-electron oxidation by methylperoxyl radicals in aqueous solutions. *Int. J. Chem. Kin.*, 18: 1185-1191.

Hwang, Y., T. Matsuto, K. Hanaki, and N. Suzuki. 1995. Identification and quantitation of sulfur and nitrogen containing odorous compounds in wastewater. *Water Research*, 29: 711-718.

Hwang, Y., T. Matsuo, K. Hanaki, and N. Suzuki. 1994. Removal of odorous compounds in wastewater by using activated carbon, ozonation and aerated biofilter. *Water Research*, 28: 2309-2319.

IARC. 1974. IARC Monographs on the evaluation of the carcinogenic risks of chemicals to man. Some aromatic amines, hydrazine and related substances, N-nitroso compounds and miscellaneous alkylating agents. Vol. 4, Lyon, France.

IARC. 1978. IARC Monographs on the evaluation of the carcinogenic risks of chemicals to humans. Some N-nitroso compounds. Vol. 17, Lyon, France.

IARC. 1987. IARC Monographs on the evaluation of carcinogenic risk of chemicals to humans. Overall evaluations of carcinogenity. Supplement 7, Lyon, France.

IARC. 2000. IARC Monographs on the evaluation of carcinogenic risk of chemicals to humans. Some industrial chemicals. Vol. 77, Lyon, France.

Islam, A.K.M.N., K. Hanaki, and T. Matsuo. 1998. Fate of dissolved odorous compounds in sewage treatment plants. *Water Science and Technology*, 38: 337-344.

Jensen, D.E., P.D. Lottikar, and P.N. Magee. 1981. The in vitro methylation of DNA by microsomally activated dimethylnitrosamine and its correlation with formaldehyde production. *Carcinogenesis*, 4: 349-354.

Jobb, D.B., R.B. Hunsinger, O. Meresz, and V. Taguchi. Removal of N-nitrosodimethylamine from the Ohsweken (six nations) water supply. Final Report. 1994. Ontario Ministry of Environment and Energy.

Kalatzis, E. and J.H. Ridd. 1966. Nitrosation, diazotization and deamination.12. Kinetics of N-nitrosation of N-methylaniline. *Jour. of the Chemical Society B – Physical Organic*, 6: 529-533.

Kaplan, D.L. and A.M. Kaplan. 1985. Biodegradation of N-nitrosodimethylamine in aqueous and soil systems. *Applied and Environmental Microbiology*, 50: 1077-1086.

Karrenbrock, F., T.P. Knepper, F. Sacher, and K. Lindner, K. 1999. Development of a standardized testfilter for the determination of microbial degradability of single compounds (In German). *Vom Wasser*, 92: 361-371.

Kefer, L.K., and P.R. Roller. 1973. N-Nitrosation by nitrite ion in neutral and basic medium. *Science*, 181: 1245-1247.

Keller J.V., J.O. Leckie, P.L. McCarty. 1978. Investigation of soluble organic nitrogen compounds in municipal secondary effluent. *Journal – Water Pollution Control Federation*, 50(11): 2522-2529.

Kimoto, W., C.J. Dooley, J. Carré, and W. Fiddler. 1980. Role of strong ion exchange resins in nitrosamine formation in water. *Water Research*, 14: 869-876.

Knepper, T.P., F. Sacher, F.T. Lange, H.-J. Brauch, F. Karrenbrock, O. Roerden, and K. Lindner. 1999. Detection of polar organic substances relevant for drinking water. *Waste Management*, 19: 77-99.

Kobayashi, M., and Y.T. Tchan. 1978. Formation of dimethylnitrosamine in polluted environment and the role of photosynthetic bacteria. *Water Research*, 12: 199-201.

Kohut, K., and S.A. Andrews. 2003a. Polyelectrolyte age and N-nitrosodimethylamine formation in drinking water treatment. *Water Quality Research Jour. of Canada*, 38: 719-735.

Kohut, K., and S.A. Andrews. 2003b. Polyelectrolytes as amine precursors for N-nitrosodimethylamine in drinking water. Proc. 2003 AWWA Ann. Conf., Anaheim, Calif.

Kohut, K.D., and S.A. Andrews. 2002. N-Nitrosodimethylamine formation in drinking water due to amine-based polyelectrolyte. Proc. 2002 AWWA WQTC, Seattle, Wash.

Kolb, E., M. Haug, C. Janzowski, A. Vetter, and G. Eisenbrand. 1997. Potential nitrosamine formation and its prevention during biological denitrification of red beet juice. *Food and Chemical Toxicology*, 35: 219-224.

Kommineni, S., W.P. Ela, R.G. Arnold, S.G. Huling, B.J. Hester, and E.A. Betterton. 2003. NDMA treatment by sequential GAC adsorption and Fenton-driven destruction. *Environmental Engineering Science*, 20: 361-373.

Krull, I. S., U. Goff, A. Silvergleid, and D. Fine. 1980. N-Nitroso compound contamination in prescription and non-prescription drugs. *Arzneimittel Forschung*, 29: 870-874.

Lake, B.G., J.C. Philips, C.E. Heading, and S.D. Gangolli. 1976. Studies on the in vitro metabolism of dimethylnitrosamine by rat liver. *Toxicology*, 5: 297-309.

Landesumweltamt Brandenburg. 2003. Bildung und Nachweis von Dimethylamin. Informationen zur Altlastenbearbeitung im Land Brandenburg Nr. 2.

Large, P.J. 1971. Non-oxidative demethylation of trimethylamine N-oxide by Pseudomonas aminovorans. *FEBS Letters*, 18: 297-300.

Large, P.J., C.A. Boulton, and M.J.C. Crabbe. 1972. The reduced nicotinamide-adenine dinucleotide phosphate- and oxygen-dependent N-oxygenation of trimethylamine by Pseudomonas aminovorans. *Biochemical Jour.*, 128: 137-138.

Leach, S.A., M. Thompson, and M. Hill. 1987. Bacterially catalyzed N-nitrosation reactions and their relative importance in the human stomach. *Carcinogenesis*, 8: 1907-1912.

Lee, C., W. Choi, Y.G. Kim, and J. Yoon. 2005a. UV photolytic mechanism of N-nitrosodimethylamine in water: Dual pathways to methylamine versus dimethylamine. *Environmental Science & Technology*, 39: 2101–2106.

Lee, C., W. Choi, and J. Yoon. 2005b. UV photolytic mechanism of N-nitrosodimethylamine in water: Roles of dissolved oxygen and solution pH. *Environmental Science & Technology*, 39: 9702–9709.

Lee, C., W. Choi, and J. Yoon. 2005c. Photocatalytic degradation of N-nitrosodimethylamine: Mechanism, product distribution, and TiO_2 surface modification. *Environmental Science & Technology*, 39: 6800–6807.

Lem, W., and P. Eng. 2000a. Application of a medium pressure UV system for the treatment of N-nitrosodimethylamine (NDMA). Calgon Carbon Corporation.

Lem, W., and P. Eng. 2000b. Scale-up of a medium pressure UV system for the treatment of N-nitrosodimethylamine (NDMA) and its advantages over low pressure UV systems. Calgon Corporation.

Liang, S., J. Min, M. Davis, J. Green, and D. Remer. 2003. Use of pulsed-UV processes to destroy NDMA. *Journal AWWA*, 95: 121-131.

Lijinsky, W., E. Conrad, and R. van de Bogart. 1982. Carcinogenic nitrosamines formed by drug/nitrite interactions. *Nature*, 239: 165-167.

Loeppky, R.N., Y.T. Bao, J.Y. Bae, L. Yu, and G. Shevlin. 1994. Blocking nitrosamine formation – Understanding the chemistry of rapid nitrosation. *ACS Symposium Series*, 553: 52-65.

Lottikar, P.D., W.J. Baldy, and E.N. Dwyer. 1975. Dimethylnitrosamine demethylation by reconstituted liver microsomal cytochrome P-450 enzyme system. *Biochemical Jour.*, 152: 705-708.

Loveland, J.P., E.G. Means, and M. Rasouli. 2001. Pilot testing to inhibit N-nitrosodimethylamine (NDMA) formation during anion exchange treatment. Proc. 2001 AWWA WQTC, Nashville, Tennessee.

Lunn, G., and E.B. Sansone. 1994. Oxidation of 1,1-dimethylhydrazine (UDMH) in aqueous solution with air and hydrogen peroxide. *Chemosphere*, 29: 1577-1590.

Lunn, G., E.B. Sansone, and A.W. Andrews. 1991. Aerial oxidation of hydrazines to nitrosamines. *Environmental and Molecular Mutagenesis*, 17: 59-62.

Luo, X., T.E. Clevenger, and D. Gang. 2003. NDMA analytical method comparisons and its occurrence in Missouri. Proceedings 2003 WQTC Conference, Philadelphia, PA.

Luo, X.H., T.E. Clevenger, and B.L. Deng. 2005. Role of NOM in the formation of N-nitrosodimethylamine (NDMA) in surface waters. Abstracts of Papers of the American Chemical Society: U848, 135-ENVR Part 1.

Mackrell, J.A., and J.R.L. Walker. 1978. Biodegradation of quaternary ammonium-compounds. *International Biodeterioration Bulletin*, 14: 77-83.

Mallik, M.A.B., and K. Tesfai. 1981. Transformation of nitrosamines in soil and invitro by soil-microorganisms. *Bulletin of Environmental Contamination and Toxicology*, 27: 115-121.

Mangino, M., and R.A. Scanlan. 1985. Nitrosation of the alkaloids hordenine and gramine, potential precursors of N-nitrosodimethylamine in barley malt. *Jour. of Agricultural and Food Chemistry*, 33, 699-705.

Meiberg, J.B.M., and W. Harder. 1978. Aerobic and anaerobic metabolism of trimethylamine, dimethylamine and methylamine in hyphomicrobium-X. *Jour. of General Microbiology*, 106: 265-276.

Mezyk, S.P., W.J. Cooper, K.P. Madden, and D.M. Bartels. 2004. Free radical destruction of N-nitrosodimethylamine in water. *Environmental Science & Technology*, 38: 3161-3167.

Mills, A.L., and M. Alexander. 1976. Factors affecting dimethylnitrosamine formation in samples of soil and water. *Jour. of Environmental Quality*, 5: 437-440.
Milstein, S., and J.B. Guttenplan. 1979. Near quantitative production of molecular nitrogen from metabolism of dimethylnitrosamine. *Biochemical and Biophysical Research Communications*, 1: 342.
Mirvish, S.S. 1970. Kinetics of dimethylamine nitrosation in relation to nitrosamine carcinogenesis. *Jour. of the National Cancer Institute*, 44: 633-639.
Mirvish, S.S. 1975. Formation of N-nitroso compounds – Chemistry, kinetics, and invivo occurrence. *Toxicology and Applied Pharmacology*, 31: 325-351.
Mirvish, S.S. 1977a. N-Nitroso compounds – Their chemical and invivo formation and possible importance as environmental carcinogens. *Jour. of Toxicology and Environmental Health*, 2: 1267-1277.
Mirvish, S.S. 1977b. N-Nitroso compounds, nitrite, and nitrate – Possible implications for causation of human cancer. *Progress in Water Technology*, 8: 195-207.
Mirvish, S.S., P. Shubik, L. Wallcave, and M. Eagen. 1972. Ascorbate-nitrite reaction – Possible means of blocking formation of carcinogenic N-nitroso compounds. *Science*, 177: 65-68.
Mitch, W.A., and D.L. Sedlak. 2002a. Factors controlling nitrosamine formation during wastewater chlorination. *Water Science and Technology*, 2: 191-198.
Mitch, W.A., and D.L. Sedlak. 2002b. Formation of N-nitrosodimethylamine from dimethylamine during chlorination. *Environmental Science & Technology*, 36: 588-595.
Mitch, W.A., and D.L. Sedlak. 2004. Characterization and fate of N-nitrosodimethylamine precursors in municipal wastewater treatment plants. *Environmental Science & Technology*, 38: 1445-1454.
Mitch, W.A., J.O. Sharp, R.R. Trussell, R.L. Valentine, L. Alvarez-Cohen, and D.L. Sedlak. 2003a. N-nitrosodimethylamine (NDMA) as a drinking water contaminant: A review. *Environmental Engineering Science*, 20: 389-404.
Mitch, W.A., A. Gerecke, and D.L. Sedlak. 2003b. A N-nitrosodimethylamine (NDMA) precursor analysis for chlorination of water and wastewater. *Water Research*, 37: 3733-3741.
Mitch W.A., G.L. Oelker, E.L. Hawley, R.A. Deeb, and D.L. Sedlak. 2005. Minimization of NDMA formation during chlorine disinfection of municipal wastewater by application of pre-formed chloramines. *Environmental Engineering Science*, 22(6): 882-890.
Mochizuki, M., N. Tsutsumi, S. Hizatate, and E. Okochi. 1994. Structures of mutagens formed from Fenton-type oxidation of N-nitrosodialkylamines. *ACS Symposium Series*, 553: 328-330.
Mofidi, A.A., J.H. Min, L.S. Palencia, B.M. Coffey, S. Liang, and J.F. Green. 2002. Advanced oxidation processes and UV photolysis for treatment of drinking water. Sacramento, California, California Energy Commission.
Mosier, A.R., and S. Torbit. 1976. Synthesis and stability of dimethylnitrosamine in cattle manure. *Jour. of Environmental Quality*, 5: 465-468.
Munoz F., and C. von Sonntag. 2000. The reactions of ozone with tertiary amines including the complexing agents nitrilotriacetic acid (NTA) and ethylenediaminetetraacetic acid (EDTA) in aqueous solution. *Journal of the Chemical Society, Perkin Transactions 1*, 2: 2029-2033.
Myers, P.A., and L.J. Zatman. 1971. The metabolism of trimethylamine N-oxide by Bacillus PM6. *Biochemical Jour.*, 121: 10.

Najm, I., and R.R. Trussell. 2001. NDMA formation in water and wastewater. *Jour. AWWA*, 93(2): 92-99.

Najm, I., S. Teefy, and S.E. Barrett. 2004. NDMA formation from water treatment polymers. Proc. 2004 AWWA Ann. Conf., Orlando, FL.

Neta, P., R.E. Huie, and A.B. Ross. 1988. Rate constants for reactions of inorganic radicals in aqueous solution. *Jour. of Physical and Chemical Reference Data* 17: 1027-1284.

Neurath, G.B., M. Duenger, F.G. Pein, D. Ambrosius, and O. Schreiber. 1977. Primary and secondary amines in the human environment. *Food and Cosmetics Toxicology*, 15: 275-282.

Nikolaou, A.D., and T.D. Lekkas. 2001. The role of natural organic matter during formation of chlorination by-products: A review. *Acta Hydrochimica et Hydrobiologica*, 29: 63-77.

Odziemkowski, M., L. Gui, and R. Gillham. 2000. Reduction of N-nitrosodimethylamine with granular iron and nickel-enhanced iron. 2. Mechanistic studies. *Environmental Science & Technology*, 34: 3495-3500.

OEHHA. 2006. Office of Environmental Health Hazard Assessment: Public Health Goal for N-Nitrosodimethylamine in drinking water. http://www.oehha.ca.gov/water/phg/pdf/122206NDMAphg.pdf

Ohshima, H., and T. Kawabata. 1978. Mechanism of N-nitrosodimethylamine formation from trimethylamine and trimethylamineoxide. Pages 143-153 in E. A. Walker and L. Griciute, eds. Environmental aspects of N-nitroso compounds. Lyon: IARC Press.

Ohta, T., J. Suzuki, Y. Iwano, and S. Suzuki. 1982. Photochemical nitrosation of dimethylamine in aqueous solution containing nitrite. *Chemosphere*, 11: 797-801.

Oliver, J.E. 1979a. Nitrosamines from pesticides. *Chemtech*, 9: 366-371.

Oliver, J.E. 1979b. Volatilization of some herbicide-related nitrosamines from soils. *Jour. of Environmental Quality*, 8: 596-601.

Oliver, J.E., P.C. Kearney, and A. Kontson. 1979. Degradation of herbicide-related nitrosamines in aerobic soils. *Jour. of Agricultural and Food Chemistry*, 27: 887-891.

Pancholy, S.K. 1978. Formation of carcinogenic nitrosamines in soils. *Soil Biology & Biochemistry*, 10: 27-32.

Parkin, G.F., and P.L. McCarty. 1981. A comparison of the characteristics of soluble organic nitrogen in untreated and activated-sludge treated wastewaters. *Water Research*, 15(1): 139-149.

Pehlivanoglu-Mantas, E., E.L. Hawley, R.A. Deeb, and D.L. Sedlak. 2006. Formation of nitrosodimethylamine (NDMA) during chlorine disinfection of wastewater effluents prior to use in irrigation systems. *Water Research*, 40: 341-347.

Pehlivanoglu-Mantas, E. and D.L. Sedlak. 2006a. Wastewater-derived dissolved organic nitrogen: Analytical methods, characterization, and effects – A review. *Critical Reviews in Environmental Science and Technology*, 36: 261-285.

Pehlivanoglu-Mantas, E. and D.L. Sedlak. 2006b. The fate of wastewater-derived NDMA precursors in the aquatic environment. *Water Research*, 40: 1287-1293.

Pietsch, J. 1997. Spurenanalytische Bestimmung polarer organischer Stickstoffverbindungen und deren Verhalten im Prozeß der Trinkwasseraufbereitung. Thesis TU Dresden.

Pietsch, J., H.-J. Brauch, and E. Worch. 1999. Verhalten polarer organischer Stickstoffverbindungen im Prozeß der Trinkwasseraufbereitung. *gwf Wasser Abwasser*, 140: 51-58.

Pietsch, J., S. Hampel, W. Schmidt, H.-J. Brauch, and E. Worch. 1996. Determination of aliphatic and acyclic amines in water by gas and liquid chromatography after derivatization by chloroformats. *Fresenius Jour. Analytical Chemistry*, 355: 164-173.

Pietsch, J., W. Schmidt, F. Sacher, H.-J. Brauch, and E. Worch. 1997. Liquid chromatographic determination of polar organic nitrogen compounds and their behaviour during drinking water treatment (In German). *Vom Wasser*, 88: 119-135.

Pietsch, J., F. Sacher, W. Schmidt, and H.-J. Brauch. 2001. Polar nitrogen compounds and their behaviour in the drinking water treatment process. *Water Research*, 35: 3537-3544.

Pitter, P., and J. Chudoba. 1990. Biodegradability of organic substances in the aquatic environment. Boca Raton, Ann Arbor, Boston: CRC Press.

Poocharoen, B., J.F. Barbour, L.M. Libbey, and R.A. Scanlan. 1992. Precursors of N-nitrosodimethylamine in malted barley. 1. Determination of hordenine and gramine. *Jour. of Agricultural and Food Chemistry*, 40: 2216-2221.

Polo, J., and Y.L. Chow. 1976. Efficient photolytic degradation of nitrosamines. *Jour. of the National Cancer Institute*, 56: 997-1001.

Preussmann, R. 1984. Occurrence and exposure to N-nitroso compounds and precursors. *IARC Scientific Publications*, 57: 3-15.

Pylypiw, H.M., G.W. Harrington, R. Kozeniauskas, and D. Diaz. 1992. The formation of N-nitrosodimethylamine and N-nitrodimethylamine from the plant-growth regulator daminozide (alar). *Jour. of Food Protection*, 55: 218-219.

Quesnel, D., and G. Nakhla. 2005. Optimization of the aerobic biological treatment of thermophilically treated refractory wastewater. *Jour. Hazardous Materials B*, 125: 221-230.

Randtke, S.J., and P.L. McCarty. 1979. Removal of soluble secondary-effluent organics. *Jour. of the Environmental Engineering Division – ASCE*, 105(4): 727-743.

Rolle, I., R. Payer, and C.J. Soeder. 1971. Über die Amine einzelliger Grünalgen. I. Der Spermidingehalt von Scenedesmus acutus (276-3a) und Chlorella fusca (211-8b). *Archiv für Mikrobiologie*, 77: 185-195.

Sacher, F., S. Lenz, and H.-J. Brauch. 1997. Analysis of primary and secondary aliphatic amines in waste water and surface water by gas chromatography-mass spectrometry after derivatization with 2,4-dinitrofluorobenzene or benzenesulfonyl chloride. *Jour. of Chromatography A*, 764: 85-93.

Sacher, F., F. Karrenbrock, Th.P. Knepper, and K. Lindner. 2001. Untersuchung der Adsorbierbarkeit von organischen Einzelstoffen als ein Kriterium ihrer Trinkwasserrelevanz. *Vom Wasser*, 96: 173-192.

Sahasrabuddhey, B., A. Jain, and K.K. Verma. 1999. Determination of ammonia and aliphatic amines in environmental aqueous samples utilizing pre-column derivatization to their phenylthioureas and high performance liquid chromatography. *Analyst*, 124: 1017-1021.

Sander, J. 1982. Endogene Nitrosaminentstehung. Pages 149-156 in DFG Deutsche Forschungsgemeinschaft, ed. Nitrat-Nitrit-Nitrosamine in Gewässern. Weinheim: Verlag Chemie.

Saunders, D.G., and J.W. Mosier. 1980. Photolysis of N-nitrosodi-normal-propylamine in water. *Jour. of Agricultural and Food Chemistry*, 28: 315-319.

Saunders, D.G., J.W. Mosier, J.E. Gray, and A. Loh. 1979. Distribution, movement, and dissipation of N-nitrosodipropylamine in soil. *Jour. of Agricultural and Food Chemistry*, 27: 584-589.

Schreiber, I.M., and W.A. Mitch. 2005. Influence of the order of reagent addition on NDMA formation during chloramination. *Environmental Science and Technology*, 39: 3811-3818.

Schreiber I.M., and W.A. Mitch. 2006a. Occurrence and fate of nitrosamines and nitrosamine precursors in wastewater-impacted surface waters using boron as a conservative tracer. *Environmental Science & Technology*, 40: 3203-3210.

Schreiber I.M., and W.A. Mitch. 2006b. Nitrosamine formation pathway revisited: The importance of dichloramine and dissolved oxygen. *Environmental Science & Technology*, 40: 6007-6014.

Sedlak, D.L., R.A. Debb, E.L. Hawley, W.A. Mitch, T.D. Durbin, S. Mowbray, and S. Carr. 2005. Sources and fate of nitrosodimethylamine and its precursors in municipal wastewater treatment plants. *Water Environment Research*, 77: 32-39.

Sen, N.P., B.A. Donaldson, and C. Charbonneau. 1975. Formation of nitrosodimethylamine from the interaction of certain pesticides and nitrite. *IARC Scientific Publications*, 9: 75-79.

Sharma, V.K., F. Kazama, H. Jiangyong, and A.K. Ray. 2005. Ferrates (iron(VI) and iron(V)): Environmentally friendly oxidants and disinfectants. *J. Water Health*, 3: 45-58.

Sharp, J.O., T.K. Wood, and L. Alvarez-Cohen. 2005. Aerobic biodegradation of N-nitrosodimethylamine (NDMA) by axenic bacterial strains. *Biotechnology and Bioengineering*, 89: 608-618.

Sharpless, C.M., and K.G. Linden. 2003. Experimental and model comparisons of low- and medium-pressure Hg lamps for the direct and H_2O_2 assisted UV photodegradation of N-nitrosodimethylamine in simulated drinking water. *Environmental Science & Technology*, 37: 1933-1940.

Siddiqui, M., and K.-Z. Atasi. 2004. N-Nitrosodimethylamine: A disinfectant byproduct and its occurrence in wastewater. *Water Environment Research*, 76(4): 316-326.

Simenhoff, M., J.J. Saukkonen, J.F. Burke, L.G. Wesson, and R.W. Schaedler. 1976. Amine metabolism and the small bowel in uraemia. *Lancet* 7990: 818-821.

Singer, S.S., W. Lijinsky, and G.M. Singer. 1978. Transnitrosation: An important aspect of the chemistry of aliphatic nitrosamines. Pages 175-181 in E. A. Walker and L. Griciute, eds. Environmental aspects of N-nitroso compounds. Lyon: IARC Press.

Slack, P.T. 1980. Nitrosamine precursors in malt. *Brauwissenschaft*, 33: 328.

Smith, T.A. 1971. The occurrence, metabolism and functions of amines in plants. *Biological Reviews of the Cambridge Philosophical Society*, 46: 201-241.

Sontheimer, H., B.R. Frick, J. Fettig, G. Hörner, C. Hubele, and G. Zimmer. 1988. Activated carbon for water treatment. DVGW-Forschungsstelle am Engler-Bunte-Institut der Universität Karlsruhe (TH), Karlsruhe.

Soroushian, F., Y. Shen, M. Patel, and M. Wehner. 2001. Evaluation and pilot testing of advanced treatment processes for NDMA removal and reformation prevention. Annual Conference Proceedings 2001. American Water Works Association.

Stefan, M.I., and J.R. Bolton. 2002. UV direct photolysis of N-nitrosodimethylamine (NDMA): Kinetic and product study. *Helvetica Chimica Acta*, 85: 1416-1426.

Stefan, M.I., M. Siddiqui, K. Atasi, and K.G. Linden. 2002. Impact of water quality characteristics on the kinetic parameters of NDMA photo-degradation. Proceedings of American Water Works Association Water Quality Technology Conference.

Stiborova, M., H.H. Schmeiser, and E. Frei. 2000. Oxidation of xenobiotics by plant microsomes, a reconstituted cytochrome P450 system and peroxidase: A comparative study. *Phytochemistry*, 54: 353-362.

Tabak, H.H., S.A. Quave, C.I. Mashni, and E.F. Barth. 1981. Biodegradability studies with organic priority pollutant compounds. *Jour. Water Pollution Control Federation*, 53: 1503-1518.

Tannenbaum, S., J.S. Wishnok, J.S. Hovis, and W.W. Bishop. 1978. N-Nitroso compounds from the reaction of primary amines with nitrite and thiocyanate. Pages 155-159 in E. A. Walker and L. Griciute, eds. Environmental aspects of N-nitroso compounds. Lyon: IARC Press.

Tate, R.L., and M. Alexander. 1974. Formation of dimethylamine and diethylamine in soil treated with pesticides. *Soil Science*, 118: 317-321.

Tate, R.L., and M. Alexander. 1975. Stability of nitrosamines in samples of lake water, soil, and sewage. *Jour. of the National Cancer Institute*, 54: 327-330.

Tate, R.L., and M. Alexander. 1976a. Microbial formation and degradation of dimethylamine. *Applied and Environmental Microbiology*, 31: 399-403.

Tate, R.L., and M. Alexander. 1976b. Resistance of nitrosamines to microbial attack. *Jour. of Environmental Quality*, 5: 131-133.

Tricker, A., B. Pfundstein, and R. Preussmann. 1994. Nitrosatable secondary amines: Exogenous and endogenous exposure and nitrosation in vivo. *ACS Symposium Series*, 553: 93-101.

Tuazon, E.C., W.P.L. Carter, R. Atkinson, A.M. Winer, and J.N. Pitts. 1984. Atmospheric reactions of N-nitrosodimethylamine and dimethylnitramine. *Environmental Science & Technology*, 18: 49-54.

U.S. Department of Health and Human Services (2005) 11th Report on carcinogens. Public Health Service, National Toxicology Program.

van Gysel, A.B., and W. Musin. 1997. Methylamines. Pages 535-541 in Ullman's encyclopedia of technical chemistry Vol. A16.

van Rheenan, D.L. 1962. Determination of biogenic amines in faeces of normal dairy cattle. *Nature*, 193: 170-171.

Vermeer, I.T.M., D.M.F.A. Pachen, J.W. Dallinga, J.C.S. Kleinjans, and J.M.S. van Maanen. 1998. Volatile N-nitrosamine formation after intake of nitrate at the ADI level in combination with an amine-rich diet. *Environmental Health Perspectives*, 106: 459-463.

von Gunten, U. 2003. Ozonation of drinking water: Part I. Oxidation kinetics and product formation. *Water Research*, 37: 1443-1467.

Wainright, T. 1986. The chemistry of nitrosamine formation – Relevance to malting and brewing. *Jour. of the Institute of Brewing*, 92: 49-64.

Walker, E.A., B. Pignatelli, and M. Castegnaro. 1979. Catalytic effect of p-nitrosophenol on the nitrosation of diethylamine. *Jour. of Agricultural and Food Chemistry*, 27: 393-396.

Wang, X.C., and C. Lee. 1995. Decomposition of aliphatic amines and amino-acids in anoxic salt-marsh sediment. *Geochimica et Cosmochimica Acta*, 59: 1787-1797.

Weerasooriya, S.V.R. and C.B. Dissanayake. 1989. The enhanced formation of N-nitrosamines in fulvic acid mediated environment. *Toxicological and Environmental Chemistry*, 25: 57-62.

Westerhoff, P., and H. Mash. 2002. Dissolved organic nitrogen in drinking water supplies: A review. *Jour. of Water Supply: Research and Technology-AQUA*, 51: 415-448.

Westerhoff P., W. Lee, J.-P. Croué, H. Gallard, and G. Amy. 2006. Organic nitrogen in drinking water and reclaimed wastewater. Final Report. Awwa Research Foundation, Denver, CO.

Wilczak, A., A. Assadi-Rad, H. Lai, L. Hoover, J. Smith, R. Berger, F. Rodigari, J. Beland, L. Lazzelle, E. Kincannon, H. Baker, and C. Heaney. 2003. Formation of NDMA in chloraminated water coagulated with DADMAC cationic polymer. *Jour. AWWA*, 95: 94-106.

Wilczak, A., H.H. Lai, and W.A. Mitch. 2004. NDMA formation in chloraminated distribution system. Is there an effect of nitrification? – EBMUD experience. Proc. 2004 AWWA Ann. Conf., Orlando, FL.

Williams, D.L.H. 1994. Quantitative aspects of nitrosamine denitrosation. *ACS Symposium Series*, 553: 66-73.

Wink, D.A., R.W. Nims, M.F. Desrosiers, P.C. Ford, and L.K. Keefer. 1991. A kinetic investigation of intermediates formed during the Fenton reagent mediated degradation of N-nitrosodimethylamine: Evidence for an oxidative pathway not involving hydroxyl radical. *Chemical Research in Toxicology*, 4: 510-512.

Wink, D.A., R.W. Nims, J.E. Saavedra, M.F. Desrosiers, and P.C. Ford. 1994. Oxidation of alkylnitrosamines via the Fenton reagent. *ACS Symposium Series*, 553: 324-327.

Wong, J. 2001. Treatment technologies for the removal of NDMA from contaminated groundwater [Online]. Available: <http://www.grac.org/wong_NDMA.pdf>. [cited August 2, 2006].

World Health Organization. 2002. N-Nitrosodimethylamine. Concise International Chemical Assessment Document 38.

www.chemfinder.com.

Xiang, Y., J.F. Sykes, and N.R. Thomson. 1996. Optimization of remedial pumping schemes for a ground-water site with multiple contaminants. *Ground Water*, 34: 2-11.

Yamamoto, G. and R.H. Sakaji. 2003. A California Regulatory perspective on NDMA. Proc. 2003 AWWA Ann. Conf., Anaheim, Calif.

Yang, W.C., J. Gan, W.P. Liu, and R. Green. 2005. Degradation of N-nitrosodimethylamine (NDMA) in ladscape soils. *Journal of Environmental Quality*, 34: 336-341.

Yordy, J.R., and M. Alexander. 1980. Microbial metabolism of N-nitrosodiethanolamine in lake water and sewage. *Applied Environmental Microbiology*, 39: 559-565.

Yordy, J.R., and M. Alexander. 1981. Formation of N-nitrosodiethanolamine from diethanolamine in lake water and sewage. *Jour. of Environmental Quality*, 10: 266-270.

Yoshinari, T. 1995. Synthesis and degradation of dimethyl nitrosamine in the natural environment and in humans. *Progress in Industrial Microbiology*, 32: 37-50.

Yoshinari, T., and D. Shafer. 1990. Degradation of dimethyl nitrosamine by methylosinus trichosporium OB3b. *Canadian Jour. Microbiology*, 36: 834-838.

Zhu, J.H., D. Yan, J.R. Xai, L.L. Ma, and B. Shen. 2001. Attempt to adsorb N-nitrosamines in solution by use of zeolites. *Chemosphere*, 44: 949-956.

Zuppi, C., I. Messana, F. Froni, C. Rossi, L. Pennacchietti, F. Ferrari, and B. Giardina. 1997. ^{1}H NMR spectra of normal urines: Reference ranges of the major metabolites. *Clinica Chimica Acta*, 265: 85-97.

ABBREVIATIONS

AC	Activated carbon
AGR	Artificial groundwater recharge
AOP	Advanced oxidation process
CAS	Chemical Abstract Service
CDOC	Chromatographic dissolved organic carbon
Cl_2	Chlorine
DBA	Di-n-butylamine
DEA	Diethylamine
DELA	Diethanolamine
DHS	California Department of Health Services
DMA	Dimethylamine
DMAB	Dimethylaminobenzene
DMAI	2-(Dimethylaminomethyl)indole
DMAP	4-Dimethylaminoantipyrine
DMCA	Dimethylchloramine
DMTC	Dimethyldithiocarbamate
DMEA	Dimethylethanolamine
DMFA	Dimethylformamide
DOC	Dissolved organic carbon
DON	Dissolved organic nitrogen
DPA	Di-n-propylamine
EA	Ethylamine
EBCT	Empty bed contact time
EDTA	Ethylenediaminetetraacetate
EE/O	Electrical energy per order
ELA	Ethanolamine
EPA	Environmental Protection Agency
EPS	Extra cellular polymeric substances
FMOC-Cl	9-Fluorenylmethylchloroformate
FP	Formation potential
GAC	Granular activated carbon
GC	Gas chromatography
GWC	Granular wood charcoal
HMW	Higher molecular weight
HOC	Hydrophobic organic carbon
HPLC	High performance liquid chromatography

IARC	International Agency for Research on Cancer
LC	Liquid chromatography
LMW	Lower molecular weight
LOD	Limit of detection
LOQ	Limit of quantification
LPML	Low pressure mercury lamp
LVI	Large volume injection
MA	Methylamine
MCL	Maximum contaminant level
MCLG	Maximum contaminant level goal
MOE	Ontario Ministry of Environment
MOR	Morpholine
MPML	Medium pressure mercury lamp
MPN	Most probable number
MS	Mass spectrometry
MSD	Mass selective detector
NDBA	N-Nitrosodi-n-butylamine
NDcHxA	N-Nitrosodicyclohexylamine
NDEA	N-Nitrosodiethylamine
NDELA	N-Nitrosodiethanolamine
NDMA	N-Nitrosodimethylamine
NDPA	N-Nitrosodi-n-propylamine
NEMA	N-Nitrosoethylmethylamine
NFP	Nitrosamine formation potential
NHMI	N-Nitrosohexamethylene imine
NMF	N-Methylformamide
NMBA	N-Nitroso-N-methylbutylamine
NMOR	N-Nitrosomorpholine
NMPA	N-Nitrosomethylpropylamine
NOM	Natural organic matter
NPA	n-Propylamine
NPIP	N-Nitrosopiperidine
NPYR	N-Nitrosopyrrolidine
O_3	Ozone
OCD	Organic carbon detection
OEHHA	California Office of Environmental Health Hazard Assessment
OND	Organic nitrogen detection
PAC	Powdered activated carbon
$pAlCl_3$	Polyaluminum chloride
PCI	Positive chemical ionization
PHG	Public Health Goal

POC	Particulate organic carbon
polyDADMAC	Polydiallyldimethylammonium chloride
PON	Particulate organic nitrogen
PYR	Pyrrolidine
RBF	Riverbank filtration
RSSCT	Rapid small scale column test
SAC	Spectral absorption coefficient
SEC	Size exclusion chromatography
SIM	Single ion monitoring
SPE	Solid-phase extraction
SUVA	Specific Ultraviolet adsorption at 254 nanometers
TEA	Triethylamine
TELA	Triethanolamine
TMA	Trimethylamine
TOC	Total organic carbon
TON	Total organic nitrogen
UDMH	Unsymmetrical dimethylhydrazine
UV	Ultraviolet
UV_{254}	UV absorption at a wavelength of 254 nm
WTP	Water treatment plant

IWA Publishing's authorised EU representative for General Product Safety Regulations is Diane D'Arras, 15 rue Duret, 75116 Paris, France, e-mail: safety@iwap.co.uk.

Printed and bound by CPI Group (UK) Ltd, Croydon, CR0 4YY
18/09/2026
02218938-0001